I. Nippert   H. Neitzel   G. Wolff (Eds.)

The New Genetics:
From Research into Health Care –
Social and Ethical Implications for Users and Providers

Springer

Berlin
Heidelberg
New York
Barcelona
Budapest
Hong Kong
London
Milan
Paris
Singapore
Tokyo

I. Nippert   H. Neitzel   G. Wolff (Eds.)

The New Genetics: From Research into Health Care

Social and Ethical Implications for Users and Providers

With 13 Figures and 39 Tables

 Springer

Professor Dr. IRMGARD NIPPERT
Westfälische Wilhelms-Universität Münster
Institut für Humangenetik
Vesaliusweg 12–14
48149 Münster, Germany

PD Dr. HEIDEMARIE NEITZEL
Medizinische Fakultät der Humboldt Universität zu Berlin
Rudolf-Virchow-Klinikum
Augustenburger Platz 1
14059 Berlin, Germany

Professor Dr. GERHARD WOLFF
Albert-Ludwigs-Universität Freiburg
Institut für Humangenetik und Anthropologie
Breisacher Straße 33
79106 Freiburg, Germany

This workshop was funded by:
Forschungs- und Förderkonzept Humangenomforschung des Bundes-
ministeriums für Bildung, Wissenschaft, Forschung und Technologie

ISBN 3-540-65920-X Springer-Verlag Berlin Heidelberg New York

Library of Congress Cataloging-in-Publication Data
The new genetics: from research into health care: social and ethical implications for users
and providers / I. Nippert, H. Neitzel, G. Wolff (eds.). p. cm.
    Includes bibliographical references.
    ISBN 3-540-65920-X (softcover: alk. paper)
1. Medical genetics – Moral and ethical aspects Congresses. 2. Human genetics – Research
– Moral and ethical aspects Congresses. I. Nippert, Irmgard. II. Neitzel, Heidemarie. III.
Wolff, G. (Gerhard) [DNLM: 1. Genetic Screening Congresses. 2. Confidentiality. 3. Ethics
Congresses. 4. Genetics, Medical Congresses. 5. Informed Consent Congresses. QZ 50
N5318 1999]' RB 155.N492   1999   174'.25–DC21   DNLM/DLC

© Springer -Verlag Berlin Heidelberg 1999
Printed in Germany

The use of general descriptive names, registered names, trademarks, etc. in this publica-
tion does not imply, even in the absence of a specific statement, that such names are ex-
empt from the relevant protective laws and regulations and therefore free for general use.

Product liability: The publishers cannot guarantee the accuracy of any information about
dosage and application contained in this book. In every individual case the user must
check such information by consulting the relevant literature.

Cover design: d&p, D-69121 Heidelberg
Production: PRO EDIT GmbH, D-69126 Heidelberg
SPIN: 10645195   27/3136-5 4 3 2 1 0– Printed on acid-free paper

Contents

Contributors

Catenhusen, Wolf-Michael
Parliamentary State Secretary of the Federal Minister
of Education and Research
(Parlamentarischer Staatssekretär bei der Ministerin
für Bildung und Forschung),
MdB, member of the Federal Parliament
(Mitglied des Bundestages),
at that time chairman of the Committee for Research,
Technology and Technology Assessment
of the German Federal Parliament
(Vorsitzender des Ausschusses für Forschung, Technologie
 und Technikfolgenabschätzung im Deutschen Bundestag),
Bundeshaus, 53113 Bonn
Friedrich-Ebert-Str. 39, 48153 Münster
Germany

Fletcher, John C., PhD
Kornfeld Professor of Biomedical –Ethics
University of Virginia School of Medicine
Box 348 HSC, Charlottesville, VA 22908
U.S.A.

Geller, Gail, ScD
Department of Pediatrics
The Johns Hopkins Medical Institutions
550 N. Broadway, Suite 301
Baltimore, MD 21205-2004
U.S.A.

HARRIS, HILARY, MD, FRCGP
General medical practitioner,
Member of the U.K. Government Advisory Committee on Genetic Testing
Catherine Road
Bowdon, Altrincham, WA14 2TD
United Kingdom

HARRIS, RODNEY, CBE, MD, FRCP, FRCPath
em. Professor of Medical Genetics,
Project Leader Concerted Action
on Genetics Services in Europe (CAGSE),
Department of medical genetics,
St. Mary's Hospital, University of Manchester,
Hathersage Road, Manchester, M13 OJH
United Kingdom

HOLTZMAN, NEIL A., MD, MPH
Professor of Pediatrics,
The Johns Hopkins University School of Medicine,
Director,
The Johns Hopkins Medical Institutions
550 N. Broadway, Suite 511, Baltimore, MD 21205-2004
U.S.A.

KENT, ALASTAIR
Director of Genetic Interest Group
29-35 Farringdon Road, London, EC1M 3JB
United Kingdom

KETTNER, MATTHIAS, Dr.
Mitglied des Kollegiums
Kulturwissenschaftliches Institut
im Wissenschaftszentrum Nordrhein-Westfalen
(Cultural Studies Center Essen)
Goethestraße 31, 45128 Essen
Germany

KRUIP, STEFAN
Dipl. Physiker
Arbeitskreis Leben mit Mukoviszidose im Mukoviszidose e.V.
Bendenweg 101, 53121 Bonn
Germany

NIPPERT, IRMGARD, Dr. rer. soc.
Professor for Women's Health Research
Institut für Humangenetik,
Westf. Wilhelms-Universität Münster
Vesaliusweg 12–14, 48149 Münster
Germany

NEITZEL, HEIDEMARIE, Dr. rer. Nat. PD
Virchow-Klinikum,
Med. Fakultät der Humboldt Universität zu Berlin
Augustenburger Platz 1, 14059 Berlin
Germany

POORTMAN, YSBRAND
VSOP Vereniging Samenwerkende Ouder-
en Patiëntenorganisaties betrokken
bij erfelijke en/of aangeboren aandoeningen
Vredehofstraat 31, 3761 HA Soesdijk
The Netherlands

RAEBURN, JA, MD
Professor of Medical Genetics
Faculty of Medicine and Health Sciences,
Centre for Medical Genetics, University of Nottingham
City Hospital, Hucknall Road, Nottingham, NG5 1PB
U.K.

SCHMIDTKE, JÖRG, Dr. med.
Professor, Head of the department of human genetics
Institut für Humangenetik,
Medizinische Hochschule Hannover
Carl-Neuberg-Straße 1, 30625 Hannover
Germany

SINGH, JAI RUP, PhD
Professor, Head of the department of human genetics
Centre for Genetic Disorders, Guru Nanak Dev University
Amritsar 143 005
India

SPERLING, KARL, Dr.
Professor
Virchow-Klinikum,
Med. Fakultät der Humboldt-Universität zu Berlin
Augustenburger Platz 1, 13353 Berlin
Germany

WERTZ, DOROTHY C., PhD
Research Professor, Senior Scientist
Division of Social Science, Ethics and Law,
The Shriver Center
200 Trapelo Road, Waltham, MA 02254
U.S.A.

WOLFF, GERHARD, Dr. med.
Professor
Institut für Humangenetik und Anthropologie,
Albert-Ludwigs-Universität
Breisacherstr. 33, 79106 Freiburg i.Br.
Germany

Introduction:
The New Genetics: From Research into Health Care –
Social and Ethical Implications for Users and Providers

IRMGARD NIPPERT, HEIDEMARIE NEITZEL, GERHARD WOLFF

Human genome research and the application of its results in health care practices is playing a pivotal role in redefining our contemporary provision of medical care.

The ultimate goals of human genome research are the treatment, cure and eventual prevention of genetic disorders but treatment and cure lag behind the ability to detect disease or increased susceptibility to disease. Most genetic services today deliver diagnosis and counseling, effective treatment like for instance in the case of PKU is rare. As more genes are identified there is growing pressure to implement new testing programs or broaden existing programs and otherwise increase both the number of available genetic tests and the amount of genetic information. [1, 2]

The potential for generating an ever increasing swell of genetic information about individuals in medical care raises crucial questions regarding the circumstances under which genetic tests should be used, how the tests should be implemented and what uses are made of their results. The main issues concerned among others are: voluntariness of services, freedom of choices, patient autonomy, informed consent, confidentiality of genetic information, privacy, testing of minors, social discrimination and stigmatization.

There is consensus evident in various reports in Europe, and the U.S.A. [3, 4, 5, 6, 7] that the appropriateness, responsiveness and competence of clinical and preventive genetic services in regard to these issues need to be assessed, that genetic technologies, genetic practices and procedures should not be left to the vagaries of economic forces, personal and/or professional interests, fears or vulnerabilities. There is growing agreement that genetic service provision and the implementation process of new genetic testing procedures should be safeguarded by providing general principles and recommendations that are based on solid facts and at the same time recognize and respect the multifaceted aspects of a pluralistic society.

Ethical and social issues raised by genetic testing often do not have a single or unambiguous solution and well informed health professionals from different specialties as well as policy makers and representatives of lay organizations may disagree with each other. Therefor it is of growing importance to discuss ethical principles and recommendations on genetic service provision in a multidisciplinary way.

At the workshop the main issues and principles that are presently emerging as integral parts of national and international recommendations on genetic service provision such as:
- voluntary provision of services
- protection of choices
- patient autonomy
- informed consent
- nondirective counseling
- confidentiality

were discussed and the participants tried to assess how these principles are known, met or violated in practice according to the newest up-to-date research findings and to identify existing gaps in data provision, research and policy analysis.

The workshop brought together an international multidisciplinary group of well known experts including health professionals, molecular biologists, social scientists and ethicists as well as representatives of patient organizations and policy makers who presented and discussed the newest data and survey findings on selected ethical and social issues in the provision of new genetic tests.

The main scientific contributors to this meeting have been awarded grants from ELSI, ESLA, BIOMED 1 and BIOMED 2 programs as well as national grants. Most of the contributors of this workshop have served or are serving on ethic committees (i.e.: Deutsche Gesellschaft für Humangenetik, American Society of Human Genetics, ESLA Board of the European Union, Dutch Health Council's Committee on Genetics and Ethics, Nuffield Council on Bioethics, Royal College of Physicians Committee on the Ethics in Medicine) or on advisory committees (i.e.: Human Genome Organization Committee of the Ethical, Social, Legal Aspects, Royal College of Physicians, Committee on Clinical Genetics, U.K. Government's Gene Therapy Advisory Committee, Committee on Assessing Genetic Risks, Division of Health Sciences Policy, Institute of Medicine of the National Academy of Sciences, U.S.A., WHO Temporary Advisors on Ethical Issues in Medical Genetics, Enquête-Kommission "Chancen und Risiken der Gentechnologie" des 10. Deutschen Bundestages) and are responsible for developing ethical guidelines for the provision of genetic services in their countries. This group of outstanding experts contributed their specialist knowledge to the development of ethical and social issues in genetics, they provided data, data analysis and interpretation for the meeting and they helped to identify existing gaps in empirical research and policy analysis.

The participants assessed what has been learned so far from recent social and ethical impact studies funded by the U.S.A. Human Genome Project (ELSI), the

European Human Genome Programme (ESLA) and BIOMED 1, as well as nationally funded studies (Deutsche Forschungsgemeinschaft and tried to identify existing gaps in data, research and policy analysis at an international level. Emerging new professional ethical guidelines and principles on the state of the art of the provision of genetic services generated in Germany, the U.K. and by WHO were presented. Areas of consensus and/or disagreement on the above mentioned ethical issues in order to promote a shared code of practice for the use of the new genetics in health care were highlighted. Developments as well as problems that have already arisen or are likely to be posed by the spread of new genetic tests like BRCA 1/2 or Cystic Fibrosis carrier screening from research settings into health care were commented upon.

The overall objective of this workshop was to discuss how the diffusion of new genetic tests from research settings into health care is best provided to ensure reasoned decision making, to identify areas of consensus and/or disagreement among professionals and consumers in order to promote a shared code of practice for the use of the new genetics in health care.

The workshop's overall philosophy was to promote a multidisciplinary discourse on ethical and social issues in order to enhance the integration of ethics in overall genetic service provision, policies and practices. It called attention to specific ethical and social issues emerging from new genetics tests and practices to clarify these issues and to take a public stand on them.

Conceptual Framework of the Meeting

Session I addressed issues regarding the circumstances under which new genetic tests should be implemented and used with special emphasis on consumer needs and protection. New data from an ELSI-study on the implementation of BRCA 1/BRCA 2 testing in regard to what women feel they need to know in order to decide whether or not to undergo testing was presented. Principles of informed consent and nondirective counseling were discussed in conjunction with new research data as well as to the overall trends in the provision of genetic tests.

Efforts of ensure the development of safe and effective tests were discussed using the recommendations of the just at the beginning of the workshop published report of the US-Task Force on Genetic Testing. The Task Force was created by the National Institutes of Health Department of Energy Working Group on Ethical, Legal, and Social Implications (ELSI) of Human Genome Research to review genetic testing in the United States of America and to make recommendations that included not only the validity and utility of genetic tests but also their appropriate use by health care providers and consumers.

Session II addressed the need for better professional training of health professionals in the provision of genetic tests. In the near future more and more genetic tests will be ordered and interpreted by primary health care practitioners. A shift of genetic testing from university based institutions to private practice is taking place already and will probably increase over the next years. Increasingly primary care providers will be called upon to refer for tests, inform patients about the availability of tests and to interpret test results. At the workshop there was consensus among the participants that a good understanding of genetics as well a good understanding of the basic principles of the provision of genetic testing such as voluntariness, informed consent, freedom of choices, nondirective counseling, confidentiality etc. should be required for all health professionals in primary care, it was agreed upon that health professionals offering or referring to genetic tests should be trained in the ethical, legal and social issues surrounding genetic diagnosis and information. Problems associated with providing genetic information in primary care settings were discussed on the basis of new data from current DFG and BIOMED 1 funded studies.

Session III addressed issues of consensus and disagreement on the provision of genetic tests among providers themselves and among providers and consumers. New data that have not been published before from current ELSI and DFG funded studies were presented for the first time in Germany, including data on consensus and/or disagreement on topics such as: voluntary versus mandatory provision of services, confidentiality versus duties to inform relatives at genetic risk, privacy of information from institutional third parties, selective abortion, presymptomatic diagnosis, testing of children and minors.

Guidelines on ethical issues in medical genetics and the provision of genetic services that were prepared in 1996–1997 in cooperation with the World Health Organization, Geneva, were presented and their universal applicability was discussed.

Session IV addressed issues on problems and prospects of future collaboration among geneticists and genetic interest groups such as parent and patient organizations.

Parent and patient organizations play a crucial part in educating geneticists on how best to provide genetic tests and counseling in ways that are more sensitive and appropriate to a variety of needs and values prevalent among parents and patients i.e.: destigmatizing language describing genetic disorders and disabilities, presenting information about genetic conditions as fairly as possible, protecting privacy and freedom of choices, accessibility and availability of services. The participation of patient groups as well as broader public participation will be required to develop appropriate approaches in public education that will respect the widely varying personal and cultural perspectives on issues of using genetics in health care that are tolerant and respectful of individuals with genetic disorders of all kinds and of individuals at risk of developing such disorders or at risk of having children with such disorders.

Therefor the organizers of the proposed meeting strongly felt that active participation of representatives of parents and patient organizations is essential to foster further collaboration among them and geneticists.

Acknowledgement

We are deeply grateful to the contributors and to our colleagues who helped making the workshop a success. Special thanks to K. Bertmaring, H. Toennies, B. Brockmann and C. Ramel for their unrelenting readiness to help running the workshop and the cheerful way they attended to the many needs of the international workshop participants.

References

1. Proceedings of the International Workshop on Cystic Fibrosis Carrier Screening Development in Europe (1997) Nippert I (ed). Women's Health Research Series I2, Münster
2. Holtzman NA (1989) Proceed with Caution, Predicting Genetic Risks in the Recombinant DNA Era, The Johns Hopkins University Press, Baltimore, London
3. Andrews LB, Fullarton JE, Holtzman NA and Motulsky AG (1994) Assessing Genetic Risks: Implication for Health and Social Policy. Washington: National Academy Press
4. Nuffield Council of Bioethics (1993) Genetic Screening: Ethical Issues
5. King's Fund Forum Consensus statement: Screening for fetal and genetic abnormality (1987) Br Med J 295:1551–3
6. European Parliament: Committee on Energy, Research and Technology: Rapporteur: Pompidou A. On the ethical aspects of the new biomedical technologies, particularly prenatal diagnosis (1993)
7. Wertz DC, Fletcher JC, Berg K (1995) Ethical Issues in Medical Genetics, WHO Hereditary Diseases Programme, Division of Noncommunicable Diseases, Genf

Opening Address

Karl Sperling

It is my pleasure and privilege to welcome you to this international workshop on „The social and ethical implications of the new Genetics". I would first like to congratulate the organizers, Irmgard Nippert, Heidemarie Neitzel and Gerhard Wolff, that they brought together representatives from so many countries and disciplines, from the old and the new world and the so-called developing countries, from social and natural sciences as well as physicians and members of the genetic support groups.

I am especially grateful that you have chosen and accepted Berlin as a meeting place which is not self-evident considering our macabre history between 1933 and 1945. This is also the main reason that in no other country the fears about misuse of our science are so marked as in Germany and that consequently your workshop is of such importance.

I was asked to say a few words concerning our past. Please let us therefore turn back the clock for exactly 70 years to the year 1927 and let us assume that this workshop is linked to the 5th International Congress of Genetics which was held in our city at that time. Here H. J. Muller presented his famous lecture on X irradiation and the induction of X-linked lethal mutations in Drosophila, here Professor Davenport, head of the Cold Spring Harbour Eugenics Records Office, the precursor of the famous Cold Spring Habor Laboratory, gave the following address at the social evening:

> „Professor Caullery has called attention to the chromosomes upon the menu card. I note that we are both placed under a picture of a chromosome complement that is defective in one chromosome. Though this is the diagram of the chromosomes of Drosophila, it may be symbolic of those of man. Let us hope that the missing chromosome is that which carries the genes for unfriendliness, for jealous rivalry, for insufficient ideals in scientific work. Let us hope that it carries the inhibitors of care in drawing conclusions in science" [1].

I do not have to remind you that he believed in the existence of defective genes in traits such as pauperism and criminality and clearly he did not carry the inhibitors of care gene in drawing conclusions in science.

At that time my brief introduction to your workshop would have been as follows:

Ladies and Gentlemen, I am convinced that you have chosen the right time and the right place to discuss such an important topic as ethics and the new genetics. It is the right time because we are convinced that the application of our science will have profound consequences for our families and especially the society as a whole and that it means nothing less then to transfer Christian charity to the next generation just as Francis Galton has told us. Berlin is the right place because for the last hundred years it permitted, even provoked cultural and intellectual non-conformity. Under the enlightened dispensation of Frederick the Great full civic equality was granted to the Jews which constitute now about 5% of our population and greatly contribute to our intellectual life. Thus, Berlin stands as an example of tolerance and liberalism. One expression of this was the foundation of our University by Wilhelm von Humboldt in 1810 which became a model for almost all modern universities. Its primary goal was to promote basic science whereby the state acts as garant for the freedom of science. This idealistic concept led to an almost unparalleled flourishing of the sciences. Here Schleiden and Schwan developed their cell doctrine, here Rudolf Virchow formulated his cellular pathology. Here, and this is another expression of the cities liberalism, the citizens elected him, Rudolf Virchow, to the Reichstag and not Bismarck. Actually, they consistently voted for the liberal opposition and not the representatives of the monarchy. In this year the Kaiser-Wilhelm Gesellschaft, our major research organisation, founded here in Berlin the first institute of Human Genetics in Germany: the Kaiser Wilhelm Institute for Anthropology, Human Heredity and Eugenics. I am convinced that in the future, let us say in 70 years, we can bring in the harvest that we are sowing in these wonderful days. Thank you for your attention.

Now, let us return to the present time. We know that six years later, from 1933 to 1945 the darkest period in German history begun with a totalitarian political system that utilised human genetics as an instrument of criminal social policy, beginning with compulsory sterilisation of several hundred thousands individuals, followed by the euthanasia program that killed many thousand patients and culminating in the most appalling excess, the holocaust, were millions of Jews and gypsies were murdered [2]. This would not have been possible without direct and indirect support from physicians and scientists, who thereby broke fundamental ethical rules.

It must be said that scientists of the KWI for Anthropology, Human Heredity and Eugenics, had been responsible to a great degree for the racial laws passed in the Third Reich. It is clear that pressure had been exerted, that opportunists offered their services and that others were Nazis by conviction. Nonetheless, it is difficult to put them all into such simple categories: Freiherr von Verschuer for example, the last director of that institute, was a member of the Confessional Church but also mentor of Dr. Dr. Joseph Mengele. The latter was well liked by his colleagues but also the most sadistic SS physician.

But let me also add that during this period, these 12 years, no institute of genetics or human genetics was founded at any German university and that it took the successor of the Kaiser-Wilhelm-Gesellschaft, the Max Planck Society, about 50 years after the end of the Second World War to move back into the field of human genetics.

The central question, why despite our humanitarian and Christian tradition, despite the liberal views of large parts of our society, such a totalitarian system as the National Socialists' in 1933 or the Communists' in 1945 could seize almost unlimited power, is still an ongoing, controversial discussion amongst historians and out of my sphere of competence.

Clearly, history never repeats itself but it is also true that those who can learn nothing from history are doomed to repeat past errors. Is there something we can learn from our history? The euthanasia program was ceased after two years due to the combined protests of the German population and church. Was there also a realistic chance to stop the discrimination of Jews, right at the beginning? In this context, I would like to cite Alfred Kühn, the famous Biologists at the KW Institute for Biology in Berlin and successor to Richard Goldschmidt, perhaps you know that he has coined the term phenocopy, who lost his German citizenship and emigrated to California:

> „I can not ignore the agonising thought, that it might have been possible to prevent much of what happened if a group of German scientists had protested the very first time Hitler attacked freedom and justice.
>
> As racial prejudice led to the removal of the first of our colleagues, if then, with one voice and with no thought of the consequences, just 50 or 100 German scholars had immediately protested, what would have happened? If Hitler had then opened the first Concentration Camp for German scientists the central power would never have germinated and there would never have been a Munich conference. It would have been a signal.
>
> And why did so much of the German youth, our students, succumb to National Socialist ideology? Well, because in us they saw, at best, disapproving neutrality. What else could we expect of them?" [3] (translated from German).

We know that many scientists were hopeful that this nightmare would soon be over, others were ambivalent, while some prominent professors, like the philosopher Heidegger, publicly supported the regime. Soon after seizure of power the Nazis took direct control over the police force and controlled the press and started a broad campaign of indoctrination of the population that is characteristic of totalitarian systems in general. One example is the most demagogic eugenic film that you can see this evening that was regularly shown as a supporting film in the German cinemas.

The lesson we can learn from our history is, perhaps, trivial: A functioning democracy with freedom of the press is the best guarantee that these extremes will not happen again. However, as you will discuss during these two days the existing problems are more subtle and there is still the risk that a well-meant intention (and let me add, the eugenic movement was also well-meant), e.g. prevention of suffering, can unwillingly led to its reverse: discrimination and stigmatisation, especially if supported by existing laws. And let me add a last point: Just recently, last week, the 100th Physican's Medical Congress was held in our country under the guidance of president Herzog where ethical problems, especially involving our field were a major subject, but also the financial constraints of our medical care system: In his opening address the president of the German Bundesärztekammer Karsten Vilmar stated:

> „Primarily, each person is responsible for his own health and that of his family. Even in a welfare state as the Federal Republic of Germany not everybody has a claim to everything" [4] (translated from German).

That is clear, but in the whole context the statement can easily be interpreted as the words of a lobbyist for the welfare of the physicians. At least for us working in the field of medical genetics it must be self-evident that we should or better must be the lobbyist for those people seeking our advice and needing our help and the measure for our reliability will be not the word but the accordance between word and action. So, please understand that at the end of my talk, I would like to give a special welcome to the representatives of the genetic support groups. Your experience and your advice are needed that the new genetics is not used for inappropriate purposes. Thank you all for coming to Berlin, I wish you a most successful workshop.

References

1. Verhandl. des V. Internat. Kongr. f. Vererbungswissenschaft (1927) p 7
2. Motulsky AG (1987) Presidential Address. Human and Medical Genetics: Past, Present, and Future. In: Vogel F and Sperling K (eds) Human Genetics – Proceedings of the 7th Int. Congress Berlin 1986, Springer-Vlg, Berlin, Heidelberg, New York
3. Kühn A (1954) In: Wissenschaft und Freiheit, Internationale Tagung Hamburg 1953. Veranst. vom Kongress für die Freiheit der Kultur und der Universität Hamburg. Berlin.
4. Der Tagesspiegel (28.5.1997). Berlin.

Session I

The Provision of the New Genetics: In Whose Best Interest?

Americans' Attitudes toward Informed Consent for Breast Cancer Susceptibility Testing: Questions for Cross-Cultural Research

GAIL GELLER

Introduction

I am a member of a National Institutes of Health (NIH) supported Consortium of investigators who have spent the last 3 years studying the ethical and social issues related to genetic testing for cancer susceptibility. My distinguished colleagues[1] and I have a grant to study informed consent for breast cancer susceptibility testing. Since you are a European, primarily German audience, I want to emphasize that my comments reflect the values of consumers and providers in the US and may or may not have meaning in other countries.

Let me begin by summarizing the most salient facts associated with inherited predisposition to breast cancer, for those of you who may not know them. About 5–10% of breast cancer is due to inherited mutations. The inherited form appears to follow an autosomal dominant pattern of inheritance which means that (1) there is a 50% chance of transmission from generation to generation, and (2) a mutation can be passed on by both genders. In addition, the inherited form of breast cancer is usually early onset, bilateral, present in several family members and associated with an increased risk of ovarian cancer. Women from families with a mutation, who themselves have the mutation, have *up to* an 85% lifetime risk of developing breast cancer (although this is lower for BRCA2 mutations than for BRCA1 mutations). On the other hand, women from families with a mutation, who themselves do *not* have a mutation, have the same lifetime risk (1 in 9) as women in the general population.

One of the critical issues that has arisen with the availability of genetic testing is how this knowledge can be useful to high risk women. No cure is available and the mechanisms of prevention and treatment are not of proven benefit. They include intensive surveillance, which usually means more frequent mammography beginning at earlier ages, prophylactic surgery (mastectomy and/or oophorectomy) and chemoprevention (such as tamoxifen). A recent retrospective study done at the

---

[1] Barbara Bernhardt, M.S. is our genetic counselor, Teresa Doksum, M.P.H. is our data analyst, Kathy Helzlsouer, M.D., M.H.S. is an oncologist, Patti Wilcox, C.N.P. is a nurse practitioner and Neil A. Holtzman, M.D., M.P.H. is a pediatrician and geneticist.

Mayo Clinic in the US indicated a 90% reduction in risk of breast cancer for women who had prophylactic mastectomy. But this study was not done among mutation carriers. That prophylactic mastectomies are offered or performed at all in the US is an issue that I know many Europeans have a hard time understanding.

The complexity of the genetics and the uncertainties associated with the decision to be tested give rise to benefits and risks of testing, which are themselves culturally determined. With regard to medical benefits, some women say they would follow their screening regimens more rigorously if they knew they had a mutation. Otherwise, except for risk reducing surgery, there is no form of prevention or treatment. Until there is, most of the benefits and risks are psychosocial. For example, there could be increased anxiety if the test is positive, decreased anxiety if the test is negative and decreased uncertainty if the result is informative. The potential for insurance discrimination is another issue that is difficult for Europeans to relate to. Because of these benefits and risks, professional organizations have recommended that informed consent be required for testing.

The concept of informed consent is a uniquely Western, particularly North American notion. It originated in the US after the Nuremberg trials to protect the interests of human subjects in research. Now it applies to clinical practice as well as research (including but not limited to genetics). Informed consent is based on certain culturally-specific norms and values regarding individual rights which are then applied to the rights and freedoms of patients and providers in the context of medical decision-making.

There are several assumptions underlying our traditional notion of informed consent. First, providers should disclose factual information that they think is important. Second, consumers (patients) should make autonomous decisions that are free from outside influence. Third, providers should refrain from making recommendations which could potentially be coercive. In the world of genetic counseling, this last assumption is the justification for *nondirective counseling*. But maybe we rely too much on what *we* think is the factual information that consumers should have, or how *we* think decisions ought to be made. The premise of our informed consent project is that there are significant so-called „agenda" differences between consumers and providers and that an ideal informed consent process should begin by exploring the values, beliefs and preferences of both groups.

Methods

One of the major components of our study was a survey we conducted last year of female consumers, physicians and nurse practitioners from specialties that are likely to be involved in such testing, to explore the following questions:

1. What topics do consumers and providers believe are important to include in *pre-test discussions?*

Table 1. Response rates

Group	N	%
High risk women	482/587	73
Nurse practitioners	143/179	80
General surgeons	55	58
Medical oncologists	47	58
*Gynecologists	51	48
*Family practitioners	77	49
*Internists	70	38
Total physicians	300	48

2. What components of *pre-test decision-making* do consumers and providers believe are important?
3. What *post-test decisions* do consumers and providers believe should be made if the test is positive?
4. Are there discrepancies among consumers, physicians and nurse practitioners in these beliefs?

Our consumers were women at increased risk of breast cancer by virtue of family history. They are the daughters and sisters of women diagnosed with breast cancer under the age of 50. We also surveyed nurse practitioners who were identified through the Maryland Chapter of their national organization. Physicians were identified through the Masterfile of the American Medical Association. We only included nurse practitioners and physicians from specialties that might encounter women at increased risk of breast cancer. These included internal medicine, family practice, obstetrics and gynecology, oncology and general surgery. All providers were given a $10 incentive to respond.

As shown in Table 1, the response rate was 73% among high risk women, and 80% among nurse practitioners. Among physicians, response rates differed by specialty, ranging from 38% among internists to 58% among surgeons and oncologists. The asterisks represent the physician specialties that are similar to those of the nurse practitioners. For the analyses in which we compare physicians and nurse practitioners, we only included these 3 physician specialties and combined them.

The survey included the following scenario. „Imagine that you are 35 years old. Your mother developed breast cancer when she was 42 and you cared for her until she died a year later. Also imagine that your sister developed breast cancer when she was 38 and is currently being treated for it. Imagine that you also have one daughter who is 13 years old. A simple test (on your blood) is available that might be able to tell you whether or not you have a very high risk of getting breast cancer." The consumer survey asked respondents to assume they were the woman in the scenario. The provider survey asked respondents to assume that the woman in the scenario was their *patient*. All respondents were then provided with *factual information* about the test, and asked various questions about this woman's decision to be tested.

Pre-Test Discussions

Table 2 shows the five out of 15 items that consumers AND providers thought were MOST important to discuss before testing. We were reassured to see how important all three groups thought it was to discuss how the chance of breast cancer can be reduced, and the chance of developing breast cancer if the test is positive. The only difference worth noting, which is underscored, is in discussing whether cancer can occur if the test is negative. Consumers do not think this is as important as do providers.

The next several figures address some of the topics to which consumers and providers attributed LESS importance. Figure 1 refers to *psychosocial aspects of testing*, and indicates that only about one third of consumers thought it was important to discuss possible depression and anxiety if the test is positive. By contrast, more than one half of providers in both groups thought this was important. With respect to the impact of testing on the relationship with one's partner, only one fourth of women thought this was important to discuss. About one third of nurse practitioners thought this issue was important. We also observed gender differences *among* physicians, with male physicians being more concerned than female physicians about relationships with partners. Perhaps this is because they are imagining the impact on *them* if their wives were tested.

Figure 2 shows differences in importance attributed to *practical aspects of testing*. Physicians underestimate the importance of discussing how the test is done

Table 2. Consumers' and providers' attitudes toward what is <u>most</u> important to discuss

Item[a]	Percent		
	Women	NPs	MDs
How the chance of breast cancer can be reduced	93	83	82
The chance of breast cancer if the test is positive	81	77	80
Possible loss of insurance if the test is positive	70	77	66
Chance the test will be positive	66	[b]	[b]
Whether cancer can occur if the test is negative	55	74	72

[a] On a 4-point scale; A total of 15 times were presented
[b] Not asked of providers

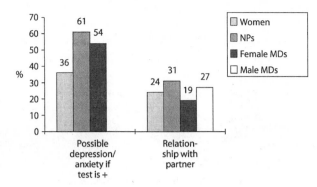

Fig. 1. Percent believing it's <u>very</u> important to discuss psychosocial aspects of testing

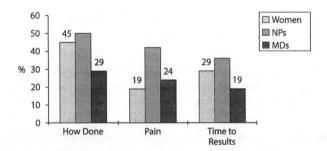

Fig. 2. Percent believing it's <u>very</u> important to discuss psychosocial aspects of testing

and how long it will take to get results, whereas nurse practitioners are more aligned with consumers. However, both groups of providers *overestimate* the importance women place on knowing whether the test will be painful. It appears, on the surface, that women do not seem to care very much about practical aspects of testing. However, as shown in Figure 3, we identified socioeconomic differences among consumers in the importance they attribute to discussing practical aspects of testing. Women from *lower* socioeconomic backgrounds place greater importance on knowing how the test will be done and how long it will take to get results than do women from higher socioeconomic backgrounds.

Pre-Test Decision-Making

In addition to the ***content*** of pre-test discussions, we examined components of the ***process*** of pre-test decision-making. With respect to autonomy, we explored what we consider to be a measure of directive counseling. Respondents were asked if, af-

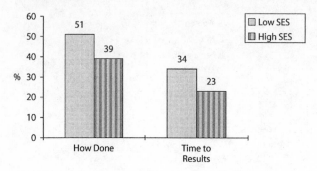

Fig. 3. SES differences among consumers in important of discussing practical issues

ter discussing the pros and cons of testing, providers should make a recommendation about whether this woman should undergo testing, or NOT make a recommendation. As shown in figure 4, over three fourths of consumers would want their providers to make a recommendation. To our surprise, physicians and nurse practitioners would be less likely to make a recommendation than women would want them to.

After the scenario, respondents were also asked their opinion about the importance of obtaining informed consent prior to such testing. Figure 5 shows that only 5% of women believe informed consent is *not* important. There is concordance between consumers' attitudes and those of nurse practitioners. By contrast, almost three times as many female physicians and five times as many male physicians thought informed consent for breast cancer susceptibility testing was not important.

Post-Test Decisions

Later in the scenario, respondents were told that the woman tested positive for a breast cancer predisposing mutation, and that her doctor told her it is not known what the best follow-up is. Respondents were then asked what post-test decisions this woman should make. Figure 6 shows the difference between providers and consumers regarding the controversial issue of prophylactic mastectomy as a follow-up option. Only 6% of women, if found to have a mutation, would want to undergo prophylactic surgery. Twice as many nurse practitioners would be likely to recommend prophylactic surgery to their high risk patients, whereas four times as many physicians would be likely to make such a recommendation.

Figure 7 shows specialty differences in physicians' tendency to recommend prophylactic mastectomy to a woman with a breast cancer susceptibility mutation. Oncologists would be more likely to recommend surgery than internists, family practitioners and gynecologists. And general surgeons would be signifi-

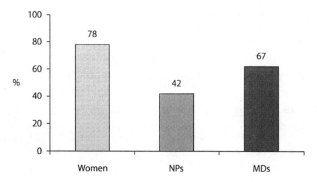

Fig. 4. Percent likely to want or make a recommendation about testing

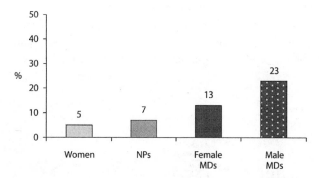

Fig. 5. Percent believing informed consent for breast cancer susceptibility testing is not important

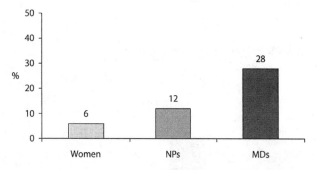

Fig. 6. Percent likely to want or recommend prophylactic mastectomy if test is +

cantly more likely than oncologists to recommend such surgery. This is not surprising, since surgeons perform these procedures.

There is no difference among consumers or providers in attitudes toward testing a 13 year old daughter of a woman with a breast cancer susceptibility mutation. As shown in Figure 8, approximately one third of all groups would think it is important. This is fairly high, and contrasts with current professional recommendations that oppose testing children.

Implications

These findings have implications for informed consent. First, women at increased risk of inherited breast cancer want information and seem to know what information is important. In order to satisfy their preferences regarding pre-test discussions, women's priorities should guide the disclosure. We should pay particular attention to socioeconomic and cultural differences in women's priorities. Sec-

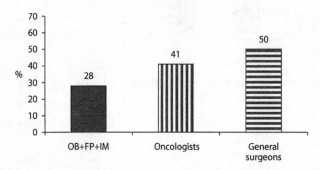

Fig. 7. Physicians likely to recommend prophylactic mastectomy by specialty

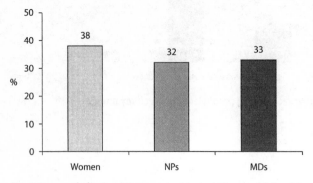

Fig. 8. Percent believing it is important to test 13 year old daughter

ond, if physicians are to be involved in education and counseling regarding breast cancer susceptibility testing, and want to adhere to professional recommendations, they should pay more attention to the importance of obtaining informed consent, to discussing practical aspects of testing and to women's concerns about prophylactic mastectomy. Third, the fact that women want their providers' recommendations suggests that we may need to rethink how we define autonomy in general, and whether nondirective counseling is appropriate in the particular context of counseling about genetic susceptibility testing. Our traditional notion of informed consent should become more of a process of shared decision-making between consumers and providers.

To the extent that informed consent is a valued goal in Europe, our experience suggests that it would be useful to conduct similar research in your countries. Some questions to explore include: (1) What role do European consumers and providers think each should play in health care decision-making? (2) What are their perceptions and beliefs about the value of, and what should be included in, informed consent for cancer susceptibility testing? (3) What do they think about the various follow-up options for high risk women? (4) Are there ethnic and class differences *within* countries, and differences *between* countries in people's attitudes toward the provider's role in clinical decision-making, informed consent for cancer susceptibility testing and follow-up options such as prophylactic mastectomy?

Nondirectiveness – Facts, Fiction, and Future Prospects

GERHARD WOLFF

Summary

Nondirectiveness is the generally required and professed standard for genetic counseling. However, studies are lacking in the field of human genetics and in other disciplines which address either the theory or practice of this type of communication in the context of genetic counseling. Moreover, there is no indication that the further development that this concept has undergone has been acknowledged in human genetics. This could be due to the historical development of genetic counseling, its inherent conflicts and often undefined goals, and the latent need of human geneticists to defend themselves against being accused of eugenic tendencies. Nondirectiveness and directiveness, however, can neither adequately describe what takes place in genetic counseling, nor can they – according to their original meaning – be used to define an ethical standard of genetic counseling.

An experiential approach is described, in which counseling is seen as a process of influence, which is wished by all the persons involved, during which activities are oriented towards the experience of the client, and which allows the counselor to communicate openly and directly with the client.

This paper critically discusses various aspects of nondirectiveness, the dealing with this concept in recent statements and investigations, the relationship between the demand of nondirectiveness and the current practice of genetic counseling, and the structural (societal) context. It is concluded that the experiential approach in genetic counseling can uphold the principle of ethics, which nondirectiveness demands and, at the same time, prevent the inevitable and unresolvable contradictions.

Introduction

According to the survey carried out by Wertz and Fletcher (1989a), more than 75% of all medical geneticists in more than 75% of the countries surveyed consider themselves committed to the principle of nondirectiveness in genetic counseling (Wertz, 1989). For many years this principle has been used in every conceivable context to indicate an ethically responsible approach to the difficulties and consequences of genetic diagnosis. This frequent usage is in striking contrast to the lack of attempts made to provide concept or substance to the term of nondirectiveness in genetic counseling, with the exception of statements to the effect that counselors should behave in a „neutral" manner and should leave decisions to their patients or clients. Both these statements are, in the final analysis, devoid of content, since it is evident that every patient/client, in the absence of direct physical or economical force, makes up his/her own mind, and „neutral" would have to be defined or conceptualized as similar as „nondirective". For example, neutrality meaning the exclusion of any subjectivity is an untenable construct which is unrealistic and has nothing to do with nondirectiveness.

Why Is Nondirective Counseling Poorly Conceptualized in Medical Genetics?

A literature research reveals that among thousands of papers dealing with genetic counseling, there are only a few which take a stand on nondirectiveness or nondirective counseling, mostly without addressing its conceptual background. One might ask why in applied human genetics, so little consideration has been given to such a central term as nondirectiveness, although it quite obviously is supposed to characterize a generally binding standard of quality of counseling in genetics, and is frequently used with this meaning. I will try to give three answers.

The **first** answer refers to the historical development of genetic counseling. Initially, genetic counseling was mainly practiced by nonphysicians to whom the paternalistic concept of the doctor-patient-relationship, with its corresponding rules on indication and advice, was unfamiliar, and who, therefore, envisaged their job as being more or less neutral transmission of information. Reed introduced the term genetic counseling in 1947 because he considered counseling in genetics as a kind of social work and wanted to divorce it from the concept of eugenics. A further clue for the introduction of nondirectiveness in genetic counseling might be seen in the fact that – at least in the U.S.A. – social workers and psychologists trained in humanistic psychology started working in the field of human genetics. They also started applying to genetic counseling the rule of nondirectiveness from humanistic psychology as founded by Carl Rogers.

All three factors mentioned – non-physicians as genetic counselors, the demarcation from the eugenics movement, and the infiltration of the humanistic psychological counseling concept – have certainly, for their part, contributed to the idea of respect for patient autonomy in genetic counseling. It is this respect for autonomy that has led to the apparent congruence between professional concepts and objectives and the interests of patients and clients. The assumption that all those involved in genetic counseling are pursuing the same objective makes it apparently superfluous to take up and conceptualize the theme of what actually occurs or should occur in genetic counseling.

Second, one can also consider nondirectiveness in genetic counseling the formulation of a problem rather than its solution. In medical genetics, the terms „illness“ and „prevention“ are poorly, if at all, defined, and in genetic counseling it is often unclear who can actually be considered a „patient“ and who, in the broadest sense of the term, is therefore to be „treated“. Genetic counselors therefore find themselves faced with the problem of defining the purpose of their intervention and the well-being of their patients and clients for whom they could justify a possibly directive influence, following the beneficence principle (helping and curing), as is widely practiced elsewhere in medicine.

In the survey of Wertz and Fletcher (1989a), medical geneticists mentioned with almost equal frequency such mutually exclusive objectives in genetic counseling as aid to individual decision-making (100%) and prevention of genetically determined diseases (97%). The improvement of the nation's state of health was considered a goal of genetic counseling by 74% and the reduction of the number of carriers of genetically determined diseases by a further 54% (Wertz, 1989, p 34). The demand for, even assertion of, „nondirectiveness“ in genetic counseling appears to be merely another way of formulating the basic problem, such as how different and sometimes conflicting values of doctor/counselor and patient/client can be respected as far as possible.

Third, the affirmation of nondirectiveness serves as a defense against attacks on the alleged harmful nature of applied human genetics. An attitude of this kind can allow no contradiction and therefore prevents critical consideration of nondirectiveness. A line of defense against open or tacit criticism of directiveness or eugenic guidance is established. Anyone practicing nondirective counseling (whatever that means) cannot be harming his/her patients/clients or pursuing morally reprehensible goals. An attitude of this kind can allow no contradiction and therefore prevents critical consideration of nondirectiveness. Used in this way, nondirectiveness degenerates into an empty slogan with no concept behind it.

A Critique of Recent Papers

Even in recent papers which explicitly refer to nondirective counseling in genetics the problems of its conceptualization have not been addressed. The document of the Task Force On Genetic Testing devotes one chapter (C.) to nondirectiveness, but without making any effort to define a concept of nondirective counseling

(Task Force On Genetic Testing 1997). Such a concept could only – if at all – be concluded from statements like those that there might be situations in which „it is difficult, and perhaps inappropriate, for providers to avoid making their preferences clear." But such an avoidance has nothing to do with nondirectiveness, and making preferences clear might even be demanded by nondirective counseling in its proper sense. The statement that „there is little opportunity for nondirectiveness in case of mandatory screening", refers to a case in which there is hardly any opportunity for any counseling at all. The statement that „providers ... should clearly state their biases but urge patients to make their own decisions" is an utterance which in its practical consequences will prove to be an attitude most likely causing confusion and helplessness in the patients. In addition, only substituting the term nondirectiveness by other terms like „nondirective stance" or „nondirective manner" (I do not know what this is) does not help to eliminate the conceptual helplessness. Likewise, the relation of nondirective counseling and informed consent would need further discussion and elucidation. Obtaining informed consent is not identical with nondirective counseling, and having obtained informed consent does not mean that nondirective counseling preceded.

The weak concept of nondirectiveness in the Task Force paper becomes evident in Principle III-13, in which the authors state that „in situations in which the provider has a bias in favor of, or opposed to, testing despite lack of convincing evidence, some believe it would be preferable for the provider to be direct and state his or her bias rather than subtly or subconsciously be directive." In saying this the authors address a „directness" which has nothing to do with „directiveness", but which – depending on the context in which it is used – can either be a powerful tool to steer the patient into the direction wanted by the provider, or a necessary and useful element of a patient oriented counseling talk. Likewise, the statement that „a recommendation by a provider may be appropriate, when the benefits outweigh the risks" can be understood as another expression of a missing concept both of directiveness and nondirectiveness. If there is a „safe and effective treatment available" (one might ask in which genetic disease this is actually the case, so that no evaluation of benefits and risks is necessary) then the accurate and understandable information about this fact is necessary and compatible with and even urged by nondirective counseling, and not a „recommendation" which the authors seems worth to be discussed as an exception of nondirectiveness, and which again prompt them to state that „people considering tests should still be encouraged to make the decision themselves". I ask myself whether the authors do not know the both bewildering and steering effect of a recommendation in connection with the hint that, of course, it's totally up to you to make your own decision, a clearly directive element, conveyed by indirect speech, which only fulfils the purpose of delivering the counselor from any responsibility.

These considerations of the context of counselor utterances lead me to the critique of another recent paper which addresses the theme of nondirectiveness

(Mitchie et al. 1997), and which prompted the American Journal of Human Genetics to issue an editorial on nondirectiveness (Bernhard 1997). The investigation of Mitchie et al. (1997) used ratings from transcripts of consultations and defined rated directiveness as advice, expressed views about or selective reinforcement of counselees' behavior, thoughts, or emotions. They defined expressions as directive apparently without considering the context, and without considering a concept of nondirectiveness and its manifestations in a counseling talk. Several of the quoted statements categorized as directive could be good examples for nondirective counseling, too. In this context I would like to refer to the work of Hartog (1996) who showed that only careful linguistic discourse analysis of precise transcripts of genetic counseling talks is able to elucidate the structure of knowledge and intentions in a given counseling. Both, the paper of Mitchie et al. and the editorial, of Bernhard disclose the same, decades old misunderstanding of nondirectiveness, namely being neutral, saying nothing which shows one's own attitudes, emotions, etc., being only an information giver and – at best – a mirror of the patient's feelings and reactions. This is the reason, why Bernhard (1997) in her editorial comes to the conclusion that „nondirective counseling is impossible to achieve", or „impossible in the context of prenatal or predictive testing". In this context, she quotes a paper of Wolff and Jung (1995), whose review of the literature allegedly provided no evidence that a nondirective approach benefits our clients. One might ask how and why it is possible that she so wrongly quotes a paper which addresses and severely criticizes the weak concept of nondirective counseling in human genetics, and which demands consequences for the education in and the practice of a further developed and more sophisticated concept of a nondirective, i.e. patient-experience oriented approach as the only one which benefits our clients. In contrast to the statement contained in the wrong quotation, another review 15 years ago had already shown that directive counseling or a counseling considered to be directive does not achieve the goals it claims it would be able to (Editorial 1982).

What Do Directiveness and Nondirectiveness Mean?

Kessler (1992) quite rightly pointed out that nondirectiveness as well as directiveness are badly defined both conceptually and operationally, and that in the realm of genetic counseling there is hardly any empirical information available about the relation between what directive or nondirective counselors say they do and what happens in practice. In his view, directiveness and nondirectiveness are merely different strategies of influence: in the first case an attempt is being made to influence behavior directly, whereas in the second it is the thought process that is being affected. Both strategies of influence are – if followed rigidly – fairly ineffective and inclined to give rise to problem situations of their own making. Kessler assesses the nondirective approach as, on the whole, more positive because it is less likely to cause conflict. Yet, he is of the opinion that in various phases of the counseling certain elements of both strategies are justified. He does not, however, provide convincing examples for using directive procedures in genetic counseling.

To learn what directiveness and nondirectiveness mean, we have to go back 50 years. According to Rogers (1942), the most important factor of the directive approach is the fact that the counselor defines the problem of the patient/client and its cause. The counselor makes proposals for further clarification and to overcome difficulties. The counselor, therefore, works on the basis of problems and results, aims at social agreement and claims the right of the capable to guide the non-capable. Rogers stressed that in this directive approach the counselor assumes great responsibility for the decisions of the client. The nondirective approach, on the other hand, is one in which the client defines the problem and selects life objectives with the counselor helping the client to find ways to achieve the stated goal(s) (Rogers, 1942, p 124).

Rogers took nondirectiveness to be an expression of humility on the part of the counselor who does not claim to have the wisdom to solve other peoples' problems but is able to assist them. Rogers' defined objective was intellectual, cognitive and emotional maturity, achieved by consolidating the feelings of self-value and of being understood. As a result, he advised counselors to act with discretion concerning decisions and evaluations.

Initially, to avoid directiveness, many guidelines were established to prevent the counselor from imposing his/her own value system, from influencing the client or from making him dependent. For example, rules forbade the counselor from stating his own opinion, answering questions, or expressing feelings directly (Gendlin, 1970). This system of rules can be understood as a reaction to the formerly widespread directive counseling. Unfortunately, it has not lost its influence on the behavior of counselors in different fields up to now including human genetics. It was and is a misunderstanding to reduce nondirectiveness to such technical rules such that the counselor is never to say or do anything which might reveal a personal experience, attitude or value and thereby act as a mirror or neutral conveyor of facts. Such a reduction in the understanding of nondirectiveness may lead to a situation whereby a patient/client is repeatedly confronted with the fact that the patient/client alone makes the decisions and the counselor acts as a participant with almost no degree of responsibility.

Rogers and others have further suggested that there are only a few crucial factors (counselor variables) which ensure that counseling is effective, such as respect, positive regard, genuineness, congruence, concretism, and empathy (for review, see Truax et al., 1965; Truax and Carkhuff, 1967). As one of the additional specific qualifications for a counselor Rogers mentioned „objectivity", which he described as controlled identification, constructive composure and emotional impartiality based on a receptive, interested attitude (Rogers, 1942, 254–255). All of these are capacities inherent in each of us, whose effectiveness can be increased through education and training and which may be harmful if insufficiently developed.

But, observing effective these counseling factors can lead to conflicts when it comes to expressing negative feelings, criticism – or, as in the case of genetic counseling – when introducing new, possibly problematic aspects in a given counseling. Merely acting on the counselor variables like rules does not necessarily lead to „good" counseling. To enable the patient/client to gain insight into his/her own situation the counselor must, if necessary, confront him/her openly and directly with facts, assessments and interpretations. However, this must be done so as to respect the personality of the client and acknowledge the unique value of the patient's/client's adaptation (Rogers, 1942, p 143 ff). Therefore, Rogers himself demanded an integration of the level of relationships and the level of task-specific interventions, and basically paved the way for the later development in humanistic psychology towards the client-centered (Rogers, 1951) and experience-oriented concepts.

Recognizing the task of this interpretation demanded already by Rogers means standing off from a „directiveness phobia", and realizing that we cannot not influence people, and that a counselor intends to influence (Litaer, 1992). In this paradigm the question of the principle of directiveness has little purpose. Instead, the questions deal with:
1. the ways and means of imposing influence,
2. the aim of such influence,
3. and the specific purpose of influence.

The contribution which a counselor can make to the counseling process is no longer formulated in a negative way (i.e. the restriction on directiveness) but rather in a positive fashion within a framework of an unavoidable, even intentional and desirable influence. The work of counseling is understood as a process to be shaped in an active way, in which an attempt is made to stimulate the unfolding of the experience-process of the patient/client. The counselor can actively take the initiative by reinforcing those factors which enable the patient/client to express his/her own experience-based adaptation to a changed situation. This kind of selective reinforcement can be called „formal", because it refers to the „form" or level of the counseling talk (cognitive, emotional etc.) (Litaer 1992). It must not be confused with selective reinforcement of content whereby the counselor attempts to force the patient/client to adopt certain attitudes or behave in a certain way which appears to the counselor to be appropriate on the basis of the objectives he himself/she herself has chosen. Up to now, no one has done an empirical, prospective study of genetic counseling the results of which allow final conclusions about the cause-effect relationship of different counseling strategies. Nevertheless, many retrospective studies from the area of genetic counseling suggest that as regards discouraging people from having children, an expressed objective on the part of directive counselors in cases of high genetic risk, directive counseling is not more effective than counseling which considers itself to be nondirective (Editorial, 1982). While the assumption of nondirectiveness in counseling claims to be able to avoid bringing this kind of substantial influence to bear, the experience-oriented approach no longer attempts to deny this influence as a matter of principle, but acknowledges its potential for achieving a more conscious contact with the manipulative potential of counseling.

What Does the Experience-Orientated Approach Means for Genetic Counseling?

With regard to genetic counseling and the task of the doctor or counselor involved, an experience-oriented approach means that the point of departure for all activities is the experience of the patient/client. Anyone consulting a doctor cannot escape the information that information (expert knowledge) and possibilities of investigation (choices of action) are available. This means that initially each individual's experience is ignored and overruled by the counselor's duty to provide certain information. Therefore, any rule committing the doctor or counselor to provide information or undertake certain measures must be carefully considered and justified on the basis of a clearly defined objective. The need for any further information and measures, and their extent must thereafter, however, be based on the experience of the client/patient who thus becomes an active subject in the counseling process.

This is in contrast to those other doctor-patient encounters, in which a generally accepted, valid goal for counseling and diagnosis exists. As a rule, in genetic counseling the objective of the individual counseling has to be elucidated in each case. This clarification must be carried out by the counselor himself/herself before the counseling, so as to avoid uncleared conflicts being conjured up in the course of the counseling. For patients/clients who do not come to counseling with an already clearly formulated objective, the possibility of clarifying this objective arises during the counseling. This should occur as soon as possible after the beginning of the sessions and should be actively encouraged by the counselor, for example by questions or inquiries about the past history of the patient/client. Should this lead to a discrepancy of objectives emerging between counselor and patient/client, these should be openly addressed in order to reach a joint counseling objective. This is not a trivial requirement since a tacit joint objective appears to be the exception rather than the rule in genetic counseling sessions. For instance, Wertz et al. (1988) showed that in only 26% of the genetic counseling cases they surveyed both the counselor and the client were aware of the major topics which the other party wished to discuss. Therefore, it seems not easy even to select a theme for genetic counseling which is jointly considered important or occurs automatically.

In addition to the responsibility for the correctness of content, the counselor is also responsible for the structure and course of the counseling. This means that he/she must introduce the necessary elements of the counseling process such as obtaining the history, investigation and interpretation of results, information about possible illnesses, and working through this information, and deal with it in a competent manner. Even the structure of the counseling must be oriented towards the experience of the patient/client and, whenever possible, should be expressed in terms of an offer from the counselor which can be accepted or rejected by the patient/client. To comply with this requirement in genetic counseling is not easy, since precisely the brevity and single nature of many counseling visits make it necessary for the counselor to prioritize various themes and problem areas for the patient/client and to structure the sequence of counseling and examination. In this regard, the counselor is required to show activity and „directness“, meaning clarity and transparency. This situation still does not relieve the counselor from the task of discussing the significance of a piece of information, a stage or result of an examination with the patient/client before each further step is undertaken. Therefore, the requirement for „concreteness“ or „directness“ must, in no case, be confused with unreflected directiveness.

Even in situations where specific questions are raised, the guideline and direction to follow, is exclusively and in every case, the experience of the patient/client. For a particular question under consideration this means, at one end of the spectrum, that it – and this is usually the case – is to be understood merely as an expression of helplessness which should indeed be taken up in the counseling. At the other end of the spectrum it can be possible to give a direct, clear answer based on the partnership established between two people acting autonomously. This latter situation will, in the case of genetic counseling, however, be exceptional because of the brevity of the contact. As a rule both „directive“ answers and „nondirective“ evasions express the counselor's latent need to control the situation (Kessler, 1992). By giving a direct answer the intention is to provide a solution to a complex decision-making situation. The question must then be asked as to how and why such a difficult situation has arisen. It could, for example, be the result of a discussion unconsciously leading towards an unclear situation which could only be resolved through a decision of the counselor. Or the patient/client with his/her question is merely reacting to latent pressure from a counselor who would like to be asked such questions in his/her capacity as an „expert in life questions “. On the other hand, evading the question in a „nondirective“ manner by indicating that it is not the counselor but the patient/client who has to make the final decision, in addition to the psychological effect of heightening the patient's/client's feelings of helplessness and not knowing what to do, has precisely the consequence of opening up the way to non-direct, concealed manipulation of decision-making.

What is necessary is in fact a basic attitude which attempts to grasp the significance of situations, findings or information for the patient/client and to interpret them. Therefore, the counselor must be flexible, which means that he/she must have at their disposal a wide range of possibilities for reaction and intervention. This requirement makes it scarcely possible to formulate a rigid „concept". What is required instead is an attitude which does not claim to be able to avoid the unavoidable influence of counseling/a counselor (as suggested for example in nondirectiveness), but rather acknowledges this influence and attempts to make use of it in a responsible way. This basic attitude and the flexibility can be practiced and learned, and a dynamic understanding of counseling can be easily adapted to the different concepts of conducting counseling sessions or to the ethics of counseling. So it is entirely possible and useful to apply some of the rigid rules of nondirectiveness in thinking about specific counseling cases or to run through the consequences of directive behavioral methods both in regard to the counseling situation and also in regard to the objectives and way in which the profession is practiced as a whole and to analyze the ethical principles being effective. An attempt should even be made to answer the question: „What would I decide? ", though only for critical reflection purposes. It is fairly straightforward to analyze one's own established values by thinking over which of the ethical principles as a counselor or – if attempting to put oneself in the situation on a trial basis – as a patient/client, one wishes to safeguard or never violate under any circumstances. This should be included as part of any initial and further training program in this field. If a counselor continually reflects over and checks in this way the basis and practice of his interventions to guide and follow his patient/client, genetic counseling will not fall into the trap of nondirectiveness, but will meet its claim to be a process desired by all of those involved, and in which the influence of the counselor is as conscious as possible.

What about the Current Practice of Genetic Counseling?

In his paper presented at the workshop „Talking Human Genetics" in April 1997 in Hamburg, Kessler stated that (p 2) „the success of the communication process in genetic counseling is no better than that found in many physician – patient interactions", and that „the counseling skills of professionals are in need of major upgrading". Professionals tend to deal with clients as information processors, decision makers, rationalists, consumers of services, and tend to approach them with a clinical detachment rather than with personal interest and involvement (pp 40 f).

From my personal experience with teaching and supervising doctors who perform genetic counseling I can say that a false or rather a non-existing concept of nondirectiveness and misunderstanding of it's task is rather the rule than the exception, and that most colleagues do not know how to follow the supposed, but undefined and unknown, mysterious rules of „correct" nondirective behavior. Many colleagues in leading positions in human genetics consider an education in counseling skills superfluous or worthless, and rather encourage young colleagues to spend another hour in the lab than gain experience in counseling. Therefore, as in other medical disciplines, active interest in patient counseling is not just beneficial to a career. In Germany, the formal education towards a so-called Facharzt für Humangenetik, a physician specialized in medical genetics including genetic counseling, which is certificated by the Deutsche Ärztekammer, the board of medical doctors, demands an education in counseling skills and in the psychological and ethical aspects of genetic counseling, as well as supervision. We developed a curriculum for this education (Wolff and Jung 1996) which has been approved by the Berufsverband Medizinische Genetik, the association of medical geneticists in Germany, and the Deutsche Gesellschaft für Humangenetik, the German society of human genetics (Gesellschaft für Humangenetik und Berufsverband Medizinische Genetik 1996). But the attendance of the workshops lends support to the assumption that it is considered a luxurious addition to rather than a conditio sine qua non for education in genetic counseling. This it at least surprising for someone who considers education in and supervision of counseling as the key to quality management in the application area of human genetics. Why not apply the same strict measures for quality management in this part of our patient oriented activities?

The practical experiences are in striking contrast to the attitudes surfacing in the answers to questions regarding nondirectiveness in a recent survey among geneticists in Germany (Nippert and Wolff, unpublished). Over 90% did not agree that a medical doctor is sufficiently educated in regard to the doctor-patient relationship so that he/she does not need a special rule like nondirectiveness. There was a more than 75% consensus that

- nondirectiveness is a necessary ethical standard for genetic counseling (91.4%);
- nondirectiveness is a sound principle for acting in a responsible way with genetic knowledge (79.1%);
- the „golden rule" that the doctor should do nothing he or she would not wish for himself or his/her next relatives, is not sufficient for genetic counseling (82.4%);
- a medical doctor should not be active in genetic counseling without a special education and qualification in counseling techniques (81.8%);
- good genetic counseling includes that the counselor addresses psychological and ethical questions without being asked to do so by the counselees (85.3%).

Therefore, the statements ascertained by surveys rather fulfill the purpose of demonstrating a responsible way of dealing with a potentially harmful technique without reflecting a practice which is in urgent need of being investigated and evaluated with adequate methods. We are in need of further investigations of the practice of counseling in genetics with methods suitable to show what is processed in the minds of counselors and counselees. One of these methods could be the detailed analysis of adequately transcribed counseling talks as shown by Hartog (1996, 1997a) and in the workshop „Talking Human Genetics" (Hartog 1997b).The results of such investigations would be an invaluable source of information about what we need for further improvement and quality management in this field. We have to learn that quality management has to and can be applied not only to laboratory measures, but also to genetic counseling, and that the only means for doing this is promoting research in this area and improving education.

What about the Structural Context of Genetic Counseling and Its Relationship to Nondirectiveness?

First, it can be stated that patient-experience oriented counseling is impossible, if the counselor is driven by a strong personal – scientific or economical interest in a specific patient decision. Therefore, counseling should be separated from any technical intervention, and – as a rule – counselor and provider of a genetic test should not be identical. Unfortunately, this essential requirement for patient-oriented counseling often is not met in both medical practice and scientific programs.

With the structural context, the institutional and societal conditions for genetic counseling should also be addressed. As already stated above, there is no place for counseling at all in the case of mandatory genetic testing. There is also little room for counseling if severe and evident economical or other societal disadvantages are impending as the consequence of certain decisions. Counselors and their clients also find themselves in a quandary, if – as in Germany – the judgement of a supreme court tries to define the goal of genetic counseling as the prevention of a handicapped child, and the birth of an handicapped child after genetic counseling of the parents as a medical failure (Wolff et al. 1995). Counselors often express the fear of legal consequences after the possible birth of a sick or handicapped child, „caused" by incomplete information or by a recommendation which was not directive enough in the sense of steering the patient's decision. Clarke (1991) takes a stand on this conflict and its structural conditions in his essay on nondirective counseling and prenatal diagnosis when he observes that nondirectiveness in connection with prenatal diagnosis „... is inevitably a sham", since in his view the mere fact of prenatal diagnosis even without directive counseling behavior establishes a directive context. According to him, nondirectiveness only serves to transfer sole moral responsibility to the parents and helps the counselors wash their hands of any responsibility („...it is their responsibility and we wash our hands of any responsibility"). A further indication of „structural" directiveness is that, for example, an abortion, after a chance detection of a sex chromosome aberration, is usually considered a successful prenatal diagnosis and not an iatrogenic catastrophe.

Clarke calls for acceptance of responsibility for the consequences of intervention, which also, among other things, reflects the potential discrimination of genetic diagnosis for those who are alive and affected and in so doing takes into account the „full price" of prenatal diagnosis and abortion.

This demand of Clarke's is not addressed only to genetic counselors but to the medical profession and even society as a whole. It is a matter of answering the question of what objectives the medical profession or society is pursuing through prenatal and predictive diagnosis. Neither nondirectiveness, whatever the term may mean, nor an approach based on the individual's experience in counseling can contribute to answering this question. The latter approach at least harbors the potential of being applied to a group situation insofar as the attitudes and experiences of those concerned (groups) are taken into account and lead the way in structuring these medical fields. Therefore, nondirectiveness understood only as a specific behavior variable of an individual counselor does not help to establish an ethically responsible way of dealing with a potentially harmful technique. Activities on a societal, political level also can be, and need to be oriented to the experience of our patients. The least we can and should do in a developed society in which in principle everybody has free access to genetic services, is, to protect those who do not want to use them, from personal and societal disadvantages.

References

Clarke A (1991) Is non-directive genetic counseling possible? Lancet II:998–1001

Editorial (1982) Directive counseling. Lancet II:368–369

Gendlin ET (1970) A short summary and some long predictions. In: Hart JT, Tomlinson TM (eds) New directions in client-centered therapy. Houghton Mifflin, Boston, pp 547–549

Gendlin ET (1974) Client-centered and experiential psychotherapy. In: Wexler DA, Rice LN (eds) Innovations in client-centered therapy. Wiley, New York, pp 211–246

Gendlin ET, Zimring F (1955) The qualities or dimensions of experiencing and their change. Counseling Center Discussion Papers I,3, Chicago

Gesellschaft für Humangenetik und Berufsverband Medizinische Genetik (1996) Zur Weiterbildung in ethischen und psychologischen Grundlagen genetischer Beratung. Med Genetik 8:237

Hartog J (1996) Das genetische Beratungsgespräch – institutionalisierte Kommunikation zwischen Experten und Nicht-Experten. Gunter Narr, Tübingen

Hartog J (1996) Das Muster genetischer Beratung – eine Diskursanalyse. Med Genetik 9:208–216

Kessler S (1989) Psychological aspects of genetic counseling: IV. A critical review of the literature dealing with education and reproduction. Am J Med Genet 34:340–353

Kessler S (1990) Current psychological issues in genetic counseling. J Psychosom Obstet Gynecol 11:5–18

Kessler S (1992) Psychological aspects of genetic counseling. VII. Thoughts on directiveness. J Genet Counseling 1:9–17

Litaer G (1992) Von „nicht-direktiv" zu „erfahrungsorientiert": Über die zentrale Bedeutung eines Kernkonzepts. In: Sachse R, Litaer G, Stiles W (Hrsg) Neue Handlungskonzepte der Klientenzentrierten Psychotherapie. Roland Asanger Verlag, Heidelberg, S 11–21

Reed S (1974) A short history of genetic counseling. Soc Biol 21:332–339

Rogers CR (1942) Counseling and psychotherapy. Houghton Mifflin, Boston

Rogers CR (1951) Client-centered Therapy. Houghton Mifflin, Boston

Sorenson JR, Swazey JP, Scotch NO (1981) Reproductive pasts, reproductive futures: Genetic counseling and its effectiveness. Birth Def Orig Art Ser XVII/4, pp 1–144

Task Force On Genetic Testing (1997) Genetic tests. C. Nondirectiveness.

Truax CB, Carkhuff RR, Kodman F (1965) Relationship between therapist-offered conditions and patient change in group psychotherapy. J Clin Psychol 21:327–329

Truax CB, Carkhuff RR (1967) Toward effective counseling and psychotherapy: Training and practice. Chicago, 1967

Wertz DC (1989) The 19-nation survey; genetics and ethics around the world. In: Wertz DC, Fletcher JC (eds) Ethics and human genetics. Springer, Berlin, Heidelberg, New York, pp 1–79

Wertz D, Fletcher JC (1989a) Ethics and human genetics. A cross-cultural perspective. Springer, Berlin, Heidelberg, New York

Wertz DC, Sorensen JR, Heeren TC (1988) Communication in health professional-lay encounters: How often does each party know what the other wants to discuss. In: Ruben BD (ed) Information and behavior, Vol 2. Transaction Books, New Brunswick,NY, pp 329–342

Wertz D, Fletcher JC, Mulvihill JJ (1990) Medical geneticists confront ethical dilemmas: cross cultural comparisons among 18 nations. Am J Hum Genet 46:1200–1213

Wolff G, Jung C (1995) Nondirectiveness and genetic counseling. J Genet Counseling 4:3–26

Wolff G, Jung C (1996) Curriculum zur Weiterbildung in ethischen und psychologischen Grundlagen genetischer Beratung. Med Genetik 8:139–142

Wolff G, Schmidtke J, Pap (1995) Das Urteil des Bundesverfassungsgerichts zum „Tübinger Fall". Med Sach 91:120–123

Predictive Genetic Tests: Destiny or Danger?

Neil A. Holtzman

Why has the human genome project generated both excitement and dread? One answer lies in James Watson's justification for the project: „We used to think our destiny was in the stars. Now we know it's in our genes." [1] Academic and commercial laboratories are already marketing genetic tests as if they can predict whether we are destined to develop certain common diseases. [2] More tests are in the research and development stage. [3] Except for rare Mendelian disorders, however, very little of our destiny is in our genes, as I will discuss first. Few tests for rare or common diseases can exclude future occurrence with certainty or, for common, complex disorders, predict occurrences with even near certainty, as I discuss second. If the public is to be protected, tests should not enter the medical marketplace until their predictive capabilities, benefits, and risks have been demonstrated. Once genetic tests enter the market, the quality of their performance must be assured. Making recommendations to implement these policies was the major accomplishment of the US Task Force on Genetic testing, as I discuss third. Despite the imperfections of most genetic tests, there are few legal restraints on insurers and employers to use the results. Nor is the confidentiality of genetic test results sufficiently protected. It is fear of discrimination and loss of privacy, as well as the inaccurate notion that our biological destiny can be foretold, that leads to dread of new genetic discoveries. In the final part of the paper, I discuss discrimination and confidentiality and recent legal efforts dealing with them.

Despite the remarkable progress made in identifying genes, the overall effects and implications remain largely unknown. Formulating policies in a time of flux due to the rapid pace of scientific discovery, is tricky business. The imprimatur of public policy, such as the endorsement of predictive genetic testing or screening when evidence is incomplete, can foster beliefs, including predestination, that eventually prove to have little basis in fact.

Is Our Destiny in Our Genes?

Claims that „our destiny" is in our genes have extended beyond disease to such complex traits as cognitive ability [4] and behaviors such as homosexuality, [5] obesity, [6] aggression, [7] and novelty-seeking. [8] The evidence is scanty at best. Few of the associations between specific genes and behavior have been confirmed let alone proven to be causal. Moreover, when single genetic factors play a significant role, they account for only a small proportion of people with the trait being examined. The allele that researchers associated with aggressive behavior has been found in only one family and may be an atypical but not unique response to subnormal cognitive ability. An intensive two year search was needed to come up with three individuals (two in one family) whose obesity appears to be due to specific genetic alterations. [9] It is highly doubtful that „obesity" genotypes in areas in which famine is endemic would be expressed in the same way as in affluent societies, which is to say that environment influences the expression of genotypes.

Although a defect in a single gene can result in a marked alteration in cognitive ability (e.g., mental retardation in phenylketonuria (PKU)) or behavior, it does not follow that these complex functions are determined by alleles at a single gene locus. The effect of knocking down one domino in a line cannot be reversed by standing one up. Many genes and environmental factors will contribute to complex behaviors. Even as we learn these contributory factors, it is doubtful that a defect in any one gene (or even two or three) will explain aberrations from what a society at a given time considers normal behavior, except in a very small proportion of cases. It is also likely that many different combinations can result in the same behavior, or even disease.

Just as PKU accounted for a very small proportion of mental retardation, so too other single-gene defects account for a small proportion of most other diseases. Specific variants of single genes confer susceptibility to common diseases, such as breast [10] and colon cancer [11] and Alzheimer disease, [12] but these alleles account for only a small proportion of all people with these diseases (usually less than 5 %). A somewhat larger proportion of the common diseases are associated with genetic polymorphisms, which occur in 1 % or more of the population. Genetic tests can seldom predict future disease with certainty regardless of whether they are intended to detect disease-causing or susceptibility-conferring alleles or polymorphisms.

Predicting Genetic Risks

The ability to predict complex behavioral traits is so nebulous that we will not discuss testing for them. For diseases inherited in accord with Mendel's laws, the finding that a healthy person possesses a disease-causing genotype will be highly predictive that future disease will occur, but seldom can all the manifestations or the severity of disease be accurately forecast. Nor can current technology detect all of the possible mutations that cause a Mendelian disorder. Over 600 mutations cause cystic fibrosis, for instance, and not all can be detected. Moreover, the severity of the lung disease in cystic fibrosis cannot be well-predicted by the disease-causing mutations that are present. [13] For the common diseases in which relatively rare alleles inherited in accord with Mendel's laws have been implicated, the appearance of disease itself (phenotype) seldom occurs in Mendelian ratios; people with positive test results will not all develop the disease. The risk of developing breast cancer, for instance, among a relatively unselected group of Ashkenazi Jewish women found to have certain susceptibility-conferring genotypes is less than 60 %. [14] For polymorphic predispositions, the risks are usually lower. Fewer than 30% of people with the apolipoprotein E4 polymorphism develop Alzheimer disease. [15] In both cases, the presence of specific alleles at other gene loci as well as environmental factors influence the actual appearance of disease. Unless very few additional gene loci are involved, and unless the risk-increasing alleles at these loci have frequencies in the polymorphic range, simultaneously testing for them (should they be discovered) will only detect a small proportion of all those with a given disease.

As the common, complex diseases, like breast cancer, usually occur in the absence of any known inherited susceptibility mutation, a negative test result (even by full-gene sequencing at over $ 2,000 a test) by no means eliminates the chance of disease. Only in families in which disease is known to be associated with a specific inherited susceptibility mutation, does the absence of that mutation in a person at risk greatly lower the chance of future disease. [16]

Although genetic tests will often reduce uncertainties of future disease, they may exacerbate previous anxieties. When the test result leads to an upward revision of the a priori risk, people will be confronted with a stronger likelihood of getting an untreatable disease. It has proved far easier for scientists to develop tests for genetic diseases than to devise effective interventions to prevent manifestations in those born affected. This therapeutic gap has already lasted over ten years for Huntington disease (HD), Duchenne muscular dystrophy (DMD) and cystic fibrosis (CF). For HD, an adult-onset dominant disorder, even some of those at risk who receive negative test results have adverse psychological reactions. A majority of people at risk of HD have, in fact, declined to be tested. [17] For all three of these disorders, prenatal diagnosis and selective pregnancy termination can be used to avoid the birth of an affected child. Among those who do not oppose abortion entirely, questions still arise about pregnancy termination for adult onset disorders, like HD, or disorders whose prognosis is improving, like CF, and for which gene therapy is already being attempted. Thus psychological and ethical issues must be considered in developing tests to predict inherited risks of future disease.

The Need to Validate Genetic Tests
and Assure Their Reliable Performance

On genetic grounds alone, therefore, many genetic tests will be imperfect predictors of disease. Yet exaggerated claims are made for genetic testing, [2,18] and tests are sometimes marketed without evidence of their predictive value. In the United States, the National Institutes of Health (NIH)-Department of Energy (DOE) Joint Working Group on Ethical, Legal, and Social Implications of Human Genome Research (hereafter the ELSI Working Group) created the Task Force on Genetic Testing in 1995 to make recommendations to ensure safe and effective genetic testing in the United States. The Task Force recently issued its final report. [19]

The Task Force would require that „protocols for the development of genetic tests that can be used predictively must receive the approval of an institutional review board (IRB)." [19, p.30] Every institution that receives federal research dollars must have an IRB. These boards were established to protect human subjects from the risks of participating in research. They must review the adequacy of safeguards of all research projects conducted at the institution. To accomplish this they must also review the scientific merit of the project. As the Federal Office responsible for IRBs has pointed out, „if a research project is so methodologically flawed that little or no reliable information will result, it is unethical to put subjects at risk or even to inconvenience them through participation." [20, p.4–1] The Task Force recommends that in reviewing protocols for genetic tests, IRBs should consider not only „the protection of human subjects involved in the study," but the adequacy of plans to collect „ data on analytic and clinical validity, and data on the test's utility for individuals who are tested." Utility refers not only to the safety and effectiveness of interventions that are undertaken in those with positive test results, but also to a reduction in psychological harm and the absence of other serious side effects. These recommendations apply to all test-development protocols in which subjects can be identified, including those conducted by biotechnology companies that do not necessarily receive Federal support for research. The Federal Government cannot compel compliance by such organizations without a change in the law. (An exception arises for companies that plan to submit their genetic tests for premarket approval to the Food and Drug Administration; the law requires them to obtain IRB approval in developing the test.)

Once data are collected under research protocols, the Task Force calls for review of the data by an independent body including consumer representatives. The results of review would influence whether the test is introduced into practice and becomes „standard of care." The Task Force does not suggest that reviewers of the data attempt to set cut points for the sensitivity and positive predictive value of tests they review; „these should vary depending on the particular test, its use, options for treatment, and other factors." [19, p.37] What is more important is that the review ensures „that the data have been appropriately collected and analyzed" and are available so that potential users can decide on the appropriateness of the test." [19, p.37] If genetic tests are marketed as tangible products, such as kits, FDA will, by law, provide the independent review. Although FDA has the authority to assess the safety and effectiveness of genetic tests marketed as services, as many genetic tests are, it has elected not to do so. [19, p.29–30]

When genetic tests actually enter the medical marketplace their performance should be subject to strict quality control. Although all laboratories performing clinical laboratory tests in the US fall under Federal control since the Clinical Laboratory Improvement Amendments (CLIA) were passed by Congress in 1988, virtually no requirements specific to genetic testing have been developed. The Task Force has recommended the creation of a genetics specialty under CLIA. (There already is a cytogenetics specialty.) Having such a specialty would facilitate establishing personnel requirements and requiring proficiency testing for laboratories performing predictive genetic tests. At the moment, participation in proficiency testing and quality control programs specifically for genetic tests is voluntary.

Genetic Discrimination

A hindrance in the development and utilization of genetic tests that can be used predictively is the fear that healthy people will be discriminated against if they are found to be at increased risk of future disease. The Task Force maintains, „No individual should be subjected to unfair discrimination by a third party on the basis of having had a genetic test or receiving an abnormal genetic test result." [19, p.15]

Health Insurance. The high costs of caring for untreatable inherited disorders sometimes leads health insurers in the United States to deny coverage to those who are known to be at risk of future disease, or charge them higher premiums, or exclude the very conditions for which they need care. [21–23] Such discrimination inhibits some people from being tested [24] and leads others to pay out of their own money for testing. Employers in the US, particularly those who pay for their employees' health care, may also be reluctant to hire those at risk of genetic disease. [25]

In the US, at least 26 states have passed laws barring health insurance discrimination on the basis of genetic testing or information. [26, 27] Employers who self-insure, however are regulated by Federal, not state, law. This gaping hole was partially closed in 1996 when Congress passed the Health Insurance Portability Act, which specifically prohibits the use of genetic information to deny group health insurance coverage when workers switch from one job to another. [28]

Employment. Under an administrative interpretation of the Americans with Disabilities Act of 1990, employers can no longer decline to hire someone based on genetic information [29], as long as the person can perform the essential functions of the job without threat to him- or herself or to others. Employers can, however, exclude from coverage disorders whose future occurrence are predicted by genetic testing after workers are employed, as long as there is an actuarial basis for doing so. [25] Stronger policies prohibiting employers from using genetic information in hiring of workers or setting their benefits and other conditions after hiring have been proposed by the National Action Plan on Breast Cancer and the ELSI Working Group. [29]

Violations of Confidentiality

Fears of discrimination as a result of testing have been compounded by worries that test results will be released or be accessible to unrelated third parties or to relatives without the explicit permission of the person tested. The Task Force recommends, „Results should be released only to those individuals for whom the test recipient has given consent for information release." [19, p.13] Participation in research may also be quelled if stored tissue samples can be used for research not originally consented by the subject or patient. [30–32] There is little disagreement that when patient identifiers are retained, informed consent is needed for uses of specimens not previously agreed to by subjects or patients. The Task Force recommends that as part of obtaining consent to participate in studies of test development „individuals must be informed of possible future uses of the specimen, whether identifiers will be retained and, if so, whether they will be recontacted." [19, p.12] The need for consent for use of specimens from which identifiers have been or will be irreversibly stripped („anonymization") is being debated. [33] Although the individual might not be harmed by anonymous use, the findings might lead to harm to the ethnic group to which the individual belongs.

To reduce the stigma that accompanies being at risk of developing an untreatable disease and to preserve the confidentiality of the very personal information regarding one's genome, legislation has been proposed in the US to protect the privacy of genetic information, including that derived from genetic testing. [29, 34] Much of this proposed legislation requires that the individual's consent be obtained for any release of genetic information, in some cases, including placement on the person's medical record. The implication of such legislation is that genetic information needs special protection; a lesser degree of confidentiality is, presumably, tolerable for other types of personal information. No consensus has been reached on this point.

Until recently, there seemed to be a strong consensus that health providers had a duty to protect the confidentiality of genetic information obtained from patients and not to convey it to relatives. Only when the patient refuses to convey information to a relative who has a high probability of suffering harm, and the harm is imminent and serious, and could be avoided, prevented, or treated as a result of conveyance of the information, might the provider invoke privilege and communicate to the relative. [35] In a recent court case in New Jersey, however, in which a woman with familial adenomatous polyposis brought suit against the estate of her deceased father's deceased physician for not warning relatives of their risk twenty years earlier, an appeals court remanded the case for trial to determine whether the duty to warn relatives was breached. [36] The law on this matter in the US is far from settled.

Conclusions

In the United States, policy makers, including the Congress and state legislatures, are increasingly aware of the potential misuses of genetic information, notably in health insurance coverage and in violations of confidentiality. Still, the concerns are often based on the belief that genetic information is capable of predicting with near certainty the future occurrence or absence of specific diseases in individuals. As I have attempted to show, seldom is there sufficient evidence to support this claim. What has lagged behind in policy making, perhaps because it challenges the „our destiny is in our genes" myth, is the need to gather evidence on the predictive capabilities of genetic tests as well as on other benefits and risks. The recommendations of the Task Force on Genetic Testing have attempted to fill this void. The question remains of whether the recommendations will be implemented. The Task Force called on the Secretary of Health and Human Services, Donna Shalala, to establish an advisory committee on genetic testing, on which the major stakeholders in genetic testing are represented, to implement its recommendations. [19, p.10] Secretary Shalala has responded by creating a department-wide working group to address the recommendations which is to report back to her periodically. The Clinical Laboratory Improvement Advisory Committee has established a genetic subcommittee to consider the Task Force recommendations pertaining to laboratory quality under CLIA. The President's National Bioethics Advisory Committee is discussing the role of IRBS, particularly in relation to genetics and genetic tests. Aware of the difficulties of dealing simultaneously with what may become a plethora of new tests, the Task Force has also recommended that the Secretary's advisory committee on genetic testing develop a scheme for defining the „stringent scrutiny" for a test, or class of tests, requires to help prioritize tests by the magnitude of policy questions they raise. [19, p.11]

Although the creation of the inter-agency working group is a major step forward it does not satisfy the Task Force's recommendation for the creation of an advisory committee on which multiple stakeholders, including consumers, the biotechnology and insurance industry, and professionals are represented. The one multiple stake-holder voice on genetics, the ELSI Working Group was disbanded early in 1997. Although a special review group recommended that the Secretary of Health and Human Services create a new Advisory Committee on Genetics and Public Policy [37], it has not been created either. At a time when genetic discoveries continue to mount, evoking dread as well as excitement, there is a need to ensure that all voices are heard in the formulation of public policy.

Task Force on Genetic Testing (1997) Promoting Safe and Effective Genetic Testing in the United States. Final Report. National Institutes of Health, Bethesda (also available on http://www.nhgri.nih.gov/elsi/tfgt_final/)

References

1. Jaroff L, Nash JM, Thompson D (1989) The gene hunt. Time, March 20:62–67
2. Holtzman NA (1995) Come to the fair. But investors should be wary at the biotech booth. Barron's, LXXV: July 10:40
3. Holtzman NA, Hilgartner S (1997) State of the art of genetic testing in the United States: Survey of biotechnology companies and nonprofit clinical laboratories and interviews of selected organizations. In: Holtzman NA, Watson MS (eds) Promoting Safe and Effective Genetic Testing in the United States. Final Report of the Task Force on Genetic Testing. National Institutes of Health, Bethesda, pp 99–124
4. Murray C, Hermstein R (1994) The Bell Curve. Free Press, New York
5. Hamer DH, Hu S, Magnuson VL, Hu N, Pattatucci AML (1993) A linkage between DNA markers on the X chromosome and male sexual orientation. Science 261:321–327
7. Zhang Y, Proenca R, Maffei M, Barone M, Leopold L, Friedman JM (1994) Positional cloning of the mouse obese gene and its human homologue. Nature 372:425–432
8. Brunner HG, Nelen MR, van Zandvoort P, Abeling NGGM, van Gennip AH, Wolters EC, et al (1993) X-Linked borderline mental retardation with prominent behavioral disturbance: Phenotype, genetic localization; and evidence for disturbed monoamine metabolism. Am J Human Genet 52:1032–1039
8. Ebstein RP, Segman R, Benjamin J, Osher Y, Nemanov L, Belmaker RH (1997) 5-HT2C (HTR2C) serotonin receptor gene polymorphism associated with the human personality trait of reward dependence: Interaction with dopamine D4 receptor (D4DR) and dopamine D3 receptor (D3DR) polymorphisms. Am J Med Genet 74:65–72
9. Leibel RL (1997) And finally, genes for human obesity. Nature Genet 16:218–220
10. Healy B (1997) BRCA genes–Bookmaking, fortunetelling, and medical care. N Engl J Med 336:1448–1449
11. Toribara NW, Sleisenger MH (1995) Screening for colorectal cancer. Current concepts. N Engl J Med 332:861–867
12. Morrison-Bogorad M, Phelps C, Buckholtz N (1997) Alzheimer disease research comes of age. The pace accelerates. JAMA 277:837–840
13. Cutting G (1996) Cystic fibrosis. In: Rimoin DL, Connor JM, Pyeritz RE (eds) Principles and Practice of Medical Genetics. Third ed. Churchill Livingstone, London
14. Struewing JP, Hartge P, Wacholder S, Baker SM, Berlin M, McAdams M, et al (1997) The risk of cancer associated with specific mutations of BRCA1 and BRCA2 among Ashkenazi Jews. N Engl J Med 336:1401–1408
15. Seshadri S, Drachman DA, Lippa CF (1995) Apolipoprotein E e4 allele and the lifetime risk of Alzheimer's disease. What physicians know, and what they should know. Arch Neurol 52:1074–1079

16. Holtzman NA (1997) Testing for genetic susceptibility to common cancers: Clinical and ethical issues. Adv Oncol 13:9–15
17. Marteau TM, Croyle R (1998) Psychological response to genetic testing. BMJ in press
18. Holtzman NA (1996) Testing for genetic susceptibility: What you see is not what you get. Accountability in Res 5:95–101
19. Task Force on Genetic Testing (1997) Promoting Safe and Effective Genetic Testing in the United States. Final Report. National Institutes of Health, Bethesda *(also available on http://www.nhgri.nih.gov/elsi/tfgt_final/)*
20. Office of Protection from Research Risks (1993) Protecting Human Research Subjects. Institutional Review Board Guidebook. US Government Printing Office, Washington DC
21. Billings PR, Kohn MA, de Cuevas M, Beckwith J, Alper JS, Natowicz MR (1992) Discrimination as a consequence of genetic testing. Am J Human Genet 50:476–482
22. Geller LN, Alper JS, Billings PR, Barash Cl, Beckwith J, Natowicz MR (1996) Individual, family, and societal dimensions of genetic discrimination: A case study analysis. Sci Eng Ethics 2:71–88
23. Lapham EV, Kozma C, Weiss JO (1996) Genetic discrimination: Perspectives of consumers. Science 274:621–624
24. Lerman C, Narod S, Schulman K, Hughes C, Gomez-Caminero A, Bonney G, et al (1996) BRCA1 testing in families with hereditary breast-ovarian cancer. A prospective study of patient decision making and outcomes. JAMA 275:1885–1892
25. Holtzman NA (1996) Medical and ethical issues in genetic screening–An academic view. Environmental Health Perspect 104:987–990
26. Rothenberg KH (1995) Genetic information and health insurance: State legislative approaches. J Law Med Ethics 23:312–319
27. Pear R (1997) States pass laws to regulate use of genetic testing. NY Times Oct 18, pp A1 –A9
28. Public Law 104–191 (1996) 110 Stat. 1936–1947
29. Rothenberg KH, Fuller B, Rothstein M, Duster T, Ellis Kahn MJ, Cunningham R, et al (1997) Genetic information and the workplace: Legislative approaches and policy challenges. Science 275:1755–1757
30. Clayton EW, Steinberg KK, Khoury MJ, Thomson E, Andrews L, Ellis Kahn MJ, et al (1995) Informed consent for genetic research on stored tissue samples. JAMA 274:1786–1792
31. American College of Medical Genetics (1995) ACMG Statement. Statement on storage and use of genetic materials. Am J Human Genet 57:1499–1500
32. American Society of Human Genetics (1996) ASHG report. Statement on informed consent for genetic research. Am J Human Genet 59:471–474
33. Holtzman NA, Andrews LB (1997) Ethical and legal issues in genetic epidemiology. Epidemiol Rev 19:163–174
34. Annas GJ, Glantz LH, Roche PA (1995) Drafting the Genetic Privacy Act: Science, policy, and practical considerations. J Law Med Ethics 23:360–366
35. Andrews L, Fullarton JE, Holtzman NA, Motulsky AG (eds) (1994) Assessing genetic risks: Implications for health and social policy. First ed. National Academy Press, Washington DC
36. Safer v. Pack (1996) 677 2D 1188 NJ
37. Marshall E (1997) Panel urges cloning ethics boards. Science 275:22

Session II

Providing the New Genetics in Primary Care: Problems and Perspectives

Genetic Services in Europe –
Primary Care Genetics Is a Priority for Health Care Systems

Rodney Harris, Hilary J. Harris, and J.A. Raeburn

Epochal changes in genetic medicine require well informed health professionals for their utilization and to ensure adequate ethical safeguards for individual patients and families. The potential magnitude of the problem has been shown by the Department of Health in British Columbia [1] who estimated that 5.5% of the population will develop genetic disease by age 25, while 60% may do so in a lifetime if one includes genetic predisposition to common disorders.

What evidence is there that medical practitioners have the necessary knowledge, skills or even awareness of medical genetics? Published evidence suggests that the opposite is true and that the present generation of doctors, in U.K. at least, has received little clinically relevant genetics at medical school [2]. Although there are attempts to rectify the situation [3], the U.K. National Confidential Enquiry into Genetic counseling (1991–1997 [4, 5, 6]) finds that obstetricians, pediatricians and internists appear to be inconsistent in providing accurate, empathic and timely genetic counseling. It cannot be assumed that specialists, who are not genetically trained, will safely and effectively use genetics for disorders even where they are a common and recurring feature of all medical practice.

The primary care team, led by the general medical practitioner (GP), is in the best situation to apply genetic information, to the individual family. This is because there are many areas of work in which the primary care team uses similar approaches to those of clinical geneticists. Table 1 lists those areas which overlap between primary care and medical genetics. Most important of all, is that both

Table 1. The common elements of genetic and primary care practice

1. Both specialities focus on the family unit
2. The use of the pedigree (or genogram) is valuable
3. Counselling is an essential part of practice
4. Confidentiality codes must be rigorously applied
5. There is long term evolving involvement with individuals

groups of specialists utilise the family as the basic unit of medical care provision. If the GP were well-trained in genetic principles and applications, then he/she is in the best situation to identify inherited problems, to instigate initial family assessments and to make appropriate referrals to the genetic centre.

Current Genetic Services in Europe

At the present time genetics services are very dependent on medical geneticists for the actual delivery of clinical and laboratory diagnosis, genetic counseling and many aspects of population genetic screening. Rapid developments are leading inevitably to increasing demand and it was uncertain whether there were sufficient well equipped genetic specialists to cope with existing and expanding work-loads. Consequently the Concerted Action on Genetics Services in Europe (CAGSE [7]) was established by the European Commission to assess the current state of specialist genetics services. This has been achieved by a survey [8] using a European network of medical geneticists working in 31 countries in the EU, many states of the former Soviet bloc and of the eastern Mediterranean region [9].

In this summary we mention only some results of the survey (Availability, Training, Recognition of the Specialty and Access), provisional recommendations and indicate on-going health services research. Figures 1&2 show the vast differences in the availability of genetic physicians and scientists in the collaborating countries.

CAGSE collaborators agreed about the requirements for an effective genetics service and about the differences between countries in what is actually abatable and accessible as is shown by the following most frequently cited concerns:

Table 2. Concerns expressed by medical geneticists in Europe (1996)

General Group	Availability	Access
Inadequate or inappropriate teaching and training	Lack in nearly half the countries of official recognition of the specialty of medical genetics	Need for comprehensive services regardless of patients' income
Need for genetic laboratory quality assessment	Inadequate numbers of trained genetics staff	Regional inequalities of provision
Problems resulting from privatization	Lack of trained counselors	Problems of funding health services generally
Inappropriate legislation	Need for a comprehensive national network of genetics centers	

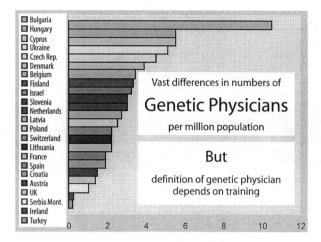

Fig. 1. Genetically trained physicians in Europe

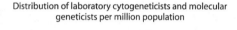

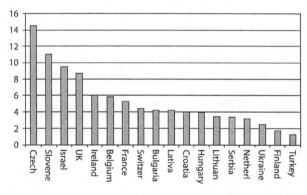

Fig. 2. Genetics laboratory scientists in Europe

As a result of the Concerted Action on Genetics Services in Europe, the following preliminary recommendations (subject to ratification by CAGSE partners in Genova May1997) have been made:

Recommendations

1. European charter for national medical genetics societies

Membership from clinical genetics, cytogenetics, molecular genetics and non-medical co-workers responsible for:

FORMAL	INFORMAL
Czech Rep.	Austria
Denmark	Belgium
Finland	Bulgaria
France	Croatia
Germany	Cyprus
Israel	Greece
Lithuania	Hungary
Netherlands	Ireland
Norway	Italy
Portugal	Latvia
Russia	Poland
Sweden	Romania
UK	Serbia Mont.
Formal training	Slovenia
includes some or all:	Spain
·Accredited training centres	Switzerland
·Agreed course contents	Turkey
·Assessment of canditates	Ukraine
·National supervision	

Fig. 3. Shows the present state of formal training for genetic physicians

UK	~1970	Latvia	Partial
Norway	1971	Belgium	No
Czech Rep.	1980	Greece	No
Finland	1982	Ireland	No
Bulgaria	1985	Portugal	No
Israel	1986	Poland	No
Netherlands	1987	Serbia Mont.	No
Russian Fed.	1988	Slovenia	No
Turkey	1990	Spain	No
Lithuania	1991	Switzerl.	No
Sweden	1991	Croatia	No data
Germany	1992	Cyprus	No data
Ukraine	1993	Hungary	No data
Austria	1994	Italy	No data
France	1995	Slovak Rep.	No data

Fig. 4. Shows the countries where medical genetics is officially recognized by the health care system

- strategic planning in co-operation with health departments and others for developing services including information for patients
- regular medical genetics research and clinical meetings
- advising on clinical and public health applications of genetic discoveries and innovations
- networking between genetics centers in the country and internationally
- reviewing data on definition and establishment of genetics personnel and facilities
- reviewing data and promoting research into access to and acceptability of genetics services
- participation in the resolution of ethical dilemmas in the context of the national characteristics

2. **Integration of regional genetics centers**

   Genetics should facilitate co-operation between the various clinical, laboratory and psychosocial aspects of the services. Wherever possible this is arranged in genetics centers serving a network of satellite clinics „hub and spoke" to:
   - provide integrated clinical, laboratory and psycho-social genetics services
   - undertake research, development and implementation
   - overcome geographical barriers by supporting satellite clinics
   - evolve appropriate links with other medical and community services to achieve optimum primary and continuing care for genetics families
   - advise and collaborate with local community based genetics services including school and other public education and become involved as the local situation requires

3. **Training and education**

   The national society and genetics centers, in collaboration with health departments, professional organizations and universities, should expedite training programs ensuring a high level of specialist and general skills including provision of accurate empathic genetic counseling in connection with genetic tests by:
   - appropriate training for specialist genetic physicians, laboratory geneticists and coworkers
   - appropriate training for primary care teams and other specialties and health care workers
   - training which includes agreed training contents, assessment of training centers, assessment of trainees and their continuing professional development after formal training has been achieved
   - clinically relevant genetics teaching for medical students

Access to Genetic Services

The ways in which patients gain access to genetics services are poorly understood and many persons may not be recognized as needing such services. When they are recognized they mostly receive genetic 'advice' from primary care, obstetricians, pediatricians, internists and other nongeneticist specialists. However, as mentioned above, the U.K. Confidential Enquiry findings suggest that specialists who have not received genetic training are not always able to facilitate the medical applications of genetics and provide adequate ethical safeguards for individual patients and families.

Primary Care and Genetics: We [10] have stressed the importance of primary care because the overall quality of genetic care is dependent on the quality of care provided by the first contact physician. This involves health professionals in making the crucial decision to reassure without further action, manage locally or refer to a specialist. Primary care may be provided by a generalist with training in family medicine and general practice, or by a specialist (obstetrician, pediatrician, internists etc.) especially in countries where there is no formal primary care training. The question arises as to whether the quality of genetic care in a country is influenced by its system of health care organization which in turn, determines which type of physician, generalist or specialist, provides first contact care.

Starfield [11] has developed a scoring system for describing the health care organization of countries in terms of their 'primary care-ness'. Measuring health care systems in this way is conceptually valuable and allows comparisons between countries. However uncertainties arise because some criteria require subjective assessments without general agreement on definitions. Health care systems are also in a state of rapid change resulting from economic pressures and other powerful forces including molecular genetics, information technology and the ethics of controlling human biology. Comparisons may also be complicated by the interaction of health care with education and social welfare systems. Poorly developed social care and widening social disparities may impact negatively upon health, irrespective of health care organization or levels of funding. With respect to genetic care, confounding variables might also include variations in the general level of public knowledge about genetic issues and the prevalence of particular disorders which heighten public awareness (e.g. sickle cell disease, thalassaemia, cystic fibrosis, Tay Sachs disease).

CAGSE has identified five criteria which appear to be the most consistent in each country for assessing primary care of patients and families with genetic problems. These are (1) the percentage of doctors who are general (not specialist) practitioners (GPs), (2) the availability of multidisciplinary primary care teams, (3) the extent to which GPs act as gatekeeper to specialist services, (4) whether patients are registered with a GP/primary care team and (5) the development of formal

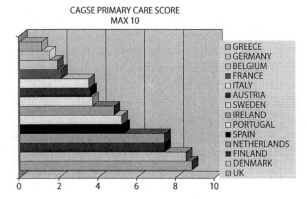

CAGSE PRIMARY CARE SCORE
MAX 10

- GREECE
- GERMANY
- BELGIUM
- FRANCE
- ITALY
- AUSTRIA
- SWEDEN
- IRELAND
- PORTUGAL
- SPAIN
- NETHERLANDS
- FINLAND
- DENMARK
- UK

Fig. 5. CAGSE initial scores for potential Primary Careness

training schemes specifically for doctors who wish to become general practitioners.

Figure 5 shows a preliminary analysis of the CAGSE study of European health services for their potential genetic „primary care-ness" using these 5 criteria. Each criterion was assigned a score between zero (denoting absence of the characteristic) and 2 (denoting a high level of development). The scores were summed (maximum 10) to yield a provisional measure of primary care for genetics (PrimGen). The analysis was based on information provided by CAGSE collaborators, by Starfield's [8] and other published data. [12,13,14] The Netherlands, Finland, Denmark and the United Kingdom score more than 7/10. Greece, Germany, Belgium and France score less than 2/10. Please note that these scores for PrimGen are provisional and do not in themselves denote any superiority or inferiority of the respective health care systems.

Health Services Research Strategy

To test the potential of genetic „primary care-ness", we are now comparing genetic service provision in countries whose health care systems differ in their primary care orientation. The aspects that we are measuring are (1) routes of referral to specialist genetic care, (2) the health care needs of referred patients (including the co-ordination of services and ongoing care); and (3) referral rates as a proportion of the estimated population prevalence rates for genetic disorders. The research involves partners in countries which have high (U.K., Denmark, Finland, Netherlands), intermediate (Sweden, Portugal, Spain) and low PrimGen scores (Greece, Germany and Belgium) as shown in figure 5.

Re-Evaluation of PrimGen

A fundamental step will be taken in the initial workshop on June 9 when general practice/primary care experts Hilary Harris (Manchester), Martin Rowland (U.K. National Primary Care R&D Centre) Barbara Starfield (Johns Hopkins) meet European medical geneticists (Harris R, U.K.; Ten Kate, Netherlands; Schmidtke, Germany; Bartsocas, Greece) to review the criteria for primary care-ness of national health care systems generally and genetic care systems in particular. This workshop will be followed (subject to successful application to EU for Shared Cost Project funding) by a workshop in each participating country at which the agreed PrimGen criteria will be applied. Each workshop will be attended by the country's leading authorities in genetics and primary care, and facilitated by a trained investigator to ensure consistency of approach across countries. A local facilitator for each country will assemble any routinely available health service statistics and workshop participants will be asked to apply the PrimGen criteria using the available statistics, supplemented, where appropriate, by their special knowledge of health care provision in the country. Discussion will focus on how perceived problems in genetic service provision may relate to the organization of health care and the implications of this for future service development and delivery.

Implementation

The findings will provide insight into how identified differences between countries in genetic service delivery may relate to the organization of health care systems. The data will be reported to EU and through CAGSE collaborators to each national government. Strong support from the European Society for Human Genetics, from national medical genetics societies, from other professional and consumer organizations will ensure adequate discussion and the implementation of concerted strategies building on existing national guidelines for training and education.

The task of providing genetic education in the primary care setting is huge. It involves not only medical professionals but the whole team, with especial attention to nurse practitioners. Unless the individual or family in which a genetic condition has been diagnosed can receive prompt, accurate and sensitive information in primary care, the whole structure of genetic services will be disadvantaged. This is particularly true if the primary care team do not appreciate the many difficult ethical areas of genetic medicine. This is a challenge for those looking to the millennium. Can we address the genetic education tasks now, so that all members of relevant health care teams receive early undergraduate teaching in genetics and follow-up training throughout their professional lives?

The full report of the Concerted Action on Genetics Services in Europe (CAGSE) was subsequently published (1997) European Journal of Human Genetics 5 (suppl 2): 1–200.

References

1. Baird PA, Anderson TW, Newcombe HB and Lowry RB (1988) Genetic disorders in children and young adults: a population study. Am J Hum Genet 42:677–93
2. Harris R, Johnston A, Harris H and Young 1 (1990) Teaching genetics to medical students. Journal of the Royal College of Physicians of London Vol 24 No 2 (April)
3. Harris R (1990) Physicians and other non-geneticists strongly favor teaching genetics to medical students in the United Kingdom. Am J Hum Genet 47:750–752
4. Report to the U.K. Department of Health from the Steering Committee of the National Confidential Enquiry into Counseling for Genetic Disorders on Down syndrome audit England and Wales, 1990–1991
5. Williamson Paula, Ponder B, Church Sarah, Fiddler Magdalen, Harris R (1996) The genetics aspects of medullary thyroid carcinoma: recognition and management. Journal of the Royal College of Physicians of London Vol 30 No 5 (September/October)
6. Williamson Paula, Dodge J, Harris R et al (1998) Genetic counseling before conception of further siblings with cystic fibrosis. Journal of the Royal College of Physicians of London in press
7. Harris R (1994) Report of ESHG Satellite Meeting: EU Concerted Action on Genetic Services in Europe (CAGSE). Eur J Hum Genet 2:300–303
8. Supplement to European Journal of Human Genetics in press
9. http://ourworld.compuserve.com/homepages/rodney_harris
10. Harris R and Harris HJ (1995) Primary care for patients at genetic risk: a priority for EC concerted action on genetics services. BMJ 311:579–580
11. Starfield B (1991) Primary care and health. A cross national comparison. JAMA 266:2268–71
12. Boerma WGW, de Jong FAJM and Mulder PH (1993) Health care and general practice across Europe. Netherlands Institute of Primary Health Care (NIVEL), Utrecht
13. Fleming D (1992) The interface between general practice and secondary care in Europe and North America. In Hospital Referrals. Edited by Martin Roland and Angela Coulter. Oxford General Practice Series 22 OUP
14. Fry John and Horder John (1994) Primary health care in an international context. Nuffield Provincial Hospitals Trust

Education in Genetics

Jörg Schmidtke

The organizers of this symposium have asked me to address the problem of how education in genetics for health professionals and consumers should best be provided. I cannot respond to this request in a scientifically satisfying way. To do so, a number of prerequisites must first be met: Different types of educational work should be tested, evaluated and compared, and both goals and measurements of success should be defined. To the best of my knowledge, no serious research in that direction has been performed in Germany.

Thus, instead of making direct suggestions I will be exploring the various circumstances, under which a need for educational measures may make itself felt.

In my experience, non-genetic primary and secondary care physicians perceive their deficits mainly in the field of *clinical* genetics, namely when trying to find out whether a disease is genetic at all, and if yes in what way. Could there be implications for the family of the affected person, including reproductive decisions and prenatal tests? The fear of litigation when such a possibility had been overlooked may contribute to this interest. Another expectation may be that a consultation with a medical geneticist could contribute to the differential diagnosis in particular cases. There is much less interest in finding out what medical geneticists are actually doing otherwise, and how they conceive their tasks themselves. It thus appears to me that a strong motivation on the part of the non-genetic professional to learn about genetics is to become aware that a genetic problem may exist at all, perhaps even to develop the skill of taking a family history and to then delegate further handling of this problem to the medical geneticist. I am not criticizing this attitude; it only reflects the fact that medicine is highly specialized and that even primary care, in Germany, is largely specialist driven.

It is a comparatively simple task, then, to encourage the primary care physician to consult with a geneticist whenever there is any doubt about the genetic nature of a disease in any given case and the implications that may follow. Problems are bound to come up, when primary care physicians over- rather than underestimate their genetic knowledge: Hybris is always a safe route towards failure.

It is the experience in many genetic counseling units, especially in urban areas, that at an increasing rate it is the affected family itself which takes the initiative to see the geneticist, and that the primary care physician's role often reduces to encourage people to do so. It is by no means rare that members of affected families, by personal experience, advice from self-support groups, and last not least from the internet have gathered an impressive amount of information, but nevertheless lack true understanding. Dealing with such cases takes special educational skills.

Thus far I had in mind situations where a clinical problem is already apparent and where the geneticist's main role is to help handling the problem of visibly increased risk in a family. The situation is quite different, however, when average rather than increased risk settings are to be considered. When it comes to genetic screening, we encounter enormous informational deficits on all parts, including primary care physicians, consumers, and especially health policy makers. A very eminent British geneticist once explained to me that in his view, genetic screening was the ideal way to educate people in genetics: „It's learning by doing!" Obviously, this colleague was convinced that genetic screening was a good thing, otherwise he wouldn't have started before people knew what it was all about. In Germany we have been more cautious.

Let me quote from the „points to consider" issued by the German Board of Medical Geneticists in March 1990, facing the possibility of a cystic fibrosis carrier screening program (Med. Genetik 2/2-3:6, 1990).

1. A cystic fibrosis carrier screening must only be offered on a voluntary basis. Any direct or indirect constraints to participate in a testing program must be excluded. Indirect constraints could already occur, if testing became part of medical routine monitoring during early pregnancy, or if non-participation would lead to disadvantages with respect to insurability.
2. In addition to informing the public at large, every individual test must be connected with an offer to explain in detail the possible consequences of its result. Such information must be accessible before a decision in favor or against is being made.
3. Cystic fibrosis heterozygote screening must be open to all. Under no circumstances must a pregnancy be made a test prerequisite.
4. Tests provision and counseling require specialized training on the part of the test provider, which needs to be formalized if testing became part of the public health care system.
5. The Board of Medical Geneticists recommends carrier screening pilot projects, should it have become technically feasible, mainly in order to test the „flanking" measures, information, education, and counseling.

In the meantime, I have become quite unhappy with this statement. I am now convinced that any carrier screening aimed at a particular disease would *principally* affect the autonomy of every individual facing the decision to take or not to take the test. The argument is simple: If medical doctors or anybody else belonging to a country's health system had made such a proposal, there must be a good reason for it. How could individuals from the general public, knowing that they have less expertise in the field, resist such an offer? In other words, even if formally voluntary, such a program would never be voluntary in reality, the authority of a professional offer could all too easily override a personal decision making process. Any educational endeavor would get caught in a double bind situation: It is virtually impossible to target a particular disease and tell people it's still their choice to use the offer or not.

It is of interest to have a closer look at the relationship between uptake rates and the style of screening offers.

A most elucidating example is given by a comparison between CF carrier screening projects performed in the former GDR and at Johns Hopkins University in Baltimore. In the GDR, CF carrier screening was outspokenly recommended to pregnant women – the uptake was 99.8% (Jung et al., Hum. Genet. 94, 19–24,1994). In Baltimore, a written invitation was sent out to HMO enrollees of child-bearing age offering an educational session prior to testing at a subsequent separate session: only 3.7% then wanted the test (Tambor et al., Am. J. Hum. Genet. 55,626–637, 1994). This pattern is practically universal: the more directive an offer is put, the higher the uptake rate will be; the more detailed the information given, and the more individual effort is placed on the part of the screenee, the lower the uptake rate. Clearly, if properly counseled, people have very little interest in pursuing very small reproductive risks. And clearly, this is the experience of every genetic counselor: when putting the CF risk of 0.05% into the perspective of the some 100-fold higher risk for any connatal disease, hardly any counselee can see the value of a test that would single out such a minor part of a much larger cumulative risk.

The conclusion to draw is that CF carrier screening programs do not appear to make much sense to the individual, if cautiously marketed. Why have carrier screening programs targeted at other diseases been so immensely successful? Let me single out one particular example: beta-thalassaemia carrier screening on Sardinia. Here, as in other Mediterranean countries, carrier screening programs have continued for decades, and wherever pursued with sufficient input, the numbers of children born with beta-thalassemia have dropped to or close to zero. This, clearly, was the aim of the programs, both when looking at individual families and at the society as a whole. As in every other recessive disease, the simple question asked was: why wait for the first affected child born, when already the first patient in a family can be avoided? But the motor driving the program was actually fuelled from a different source:

Antonio Cao, the initiator of the beta-thalassemia carrier screening on Sardinia, once explained to me: „In Sardinia the carrier screening for the prevention of thalassemia became urgent when a cure based on blood transfusions and

Desferal was introduced, which alone would have led to a hardly controllable increase in the number of affected subjects (100 new cases per year in absence of prevention). In other words, I do not think that the cure itself has been the determining factor for the beginning of the screening program, but it has certainly be a relevant "vis a tergo" for its working out."

It is quite obviously the balance of resources available for the maintenance of health on the one side and treatment of disease on the other is the major factor that will determine the fate of any genetic carrier screening program, in Sardinia and anywhere else. If the prevention of genetic disease is a declared goal of a society, then naturally such programs are eugenic in the sense that they are meant to „interfere with reproductive decisions of individuals to reach a societal goal" (a definition we owe to N.A.Holtzman, „Proceed with caution", Baltimore, 1989).

Thus far, I have limited myself to genetic carrier screening. Clearly, there are other settings which also deserve to be included under the label of genetic screening, which is usually understood as the search for genotypes in an asymptomatic population, which lead to increased risk for genetic disease in their carriers or offspring. I would like to stress, however, that the „search for genotypes" must not necessarily be performed at the level of the genetic material itself but may initially include phenotypic markers such as MSAFP levels or ultrasound abnormalities.

Prenatal genetic screening is performed here and elsewhere for decades now. It includes chromosome analyses, ultrasound and serum marker tests, and all of these measures appear to be widely accepted in the target groups. As we have seen, high acceptance rates correlate inversely with the rate of pre-test counseling, and indeed, as we all know, only a minority of pregnant women receive proper counseling before non-invasive measures are taken, and even before invasive procedure, genetic counseling is the exception rather than the rule. Women are notoriously ill-informed, and while invasive measures still require informed consent at least formally, there is little doubt that only few women really understand beforehand what is being done to them when blood is taken for a „triple-test" or a special ultrasound is applied in search for fetal abnormalities.

Cost-benefit analyses had played an important role when prenatal screening programs were initiated in Germany and elsewhere, and these were designed along the same eugenic lines of reasoning as I have just defined. Until very recently, invasive prenatal chromosome analysis was actively offered only to women over 35 years of age. Such an age limit makes sense only if limited resources are to be channeled in a most effective way, namely by maximizing detection rates.

Indeed, we are having a very difficult job to now redefine prenatal measures as being in the interest of the pregnant woman as an individual, when so little is being done to first create a proper counseling setting before tests are applied. Current surveys indicate that the relative share of genetic counseling among all measures provided by human geneticists is decreasing, not increasing (Nippert et al., Med.

Genetik 2, 188–205, 1997), and there is no evidence that public awareness of medical genetics and education in human genetics would be growing.

It is of course of great interest to investigate the reasons why individual counseling in connection with general genetic test offers is underdeveloped. Most certainly, the blame is on the part of the professionals, not the screenees. First, there aren't enough counselors. Although the exact number of medical graduates enrolled in medical genetics educational programs in Germany is not known, one can estimate that there are probably less than 200 positions devoted solely to further education in medical genetics including counseling – far too few to meet the increasing need in view of growing test offers. There is no indication that the situation will improve. On the contrary, medical faculties are trying to curb the activities of human genetics departments, as human genetics is penetrating medicine as a whole. „We all do human genetics", is a belief that many medical scholars share nowadays. But: human genetics is absorbed as a technology only, whereas genetic counseling is overlooked, belittled or ignored.

In order to summarize my line of reasoning: If screening programs are to be optimized in terms of uptake rates, education may actually be seen to be counterproductive in that it may diminish their success.

Genetic screening is the meeting point of human genetics and public health. It is my impression that we are, with so many open questions, at a very rudimentary stage of the discussion. Those in this country, whose job it is to set the stage for future health economy (such as the „Sachverständigenrat im Gesundheitswesen", Gesundheitsversorgung und Krankenversicherung 2000, Baden-Baden, 1995), appear largely ignorant of human genetics. From my experience from giving a 3-hour crash course in human genetics to students specializing in public health I know that the next generation of public health professionals will be almost as ignorant about genetics as the present one. My conclusion is: Genetic screening programs in this country will not be evaluated properly, neither economically, politically, nor ethically, they will not even be called that way, but they will take place by means of diffusion into everyday medicine. People will get used to test offers, at the individual level, aiming at late onset diseases just as they have got used to prenatal screening for chromosome analyses and will get used to demand prenatal screening for single gene disorders. Educational measures, irrespective of being mainly propagandistic or asking for caution, will be a threat to this diffusion process and will be discouraged, actively or passively, and this will be most pronounced in screening situations, where, in view of consumer autonomy, education is needed most.

However, even from the point of view of test offerers, including the test manufacturing industry, discouragement of public education is a mistake on the long run. A good businessman wants happy customers. With a society having split views over the value of (predictive) genetic testing, every single case, where someone was taken by surprise by genetic testing without proper pre-test counseling will contribute to a bad image of the whole test industry. This cannot possibly be in their interest. Where we, patients, counselees, counselors, doctors, and test manufacturers will finally meet and where our interaction is evaluated, is the field of quality assessment. Genetic education with its many facets, including pre- and post-test counseling, teaching at medical schools and in specialist and non-specialist training, is indispensable in developing optimal medical-genetic care.

Genesis, Compulsions, Acceptability, and Future of Pre-Natal Sex-Selection in India

Jai Rup Singh

India has the highest number of female Goddesses being worshipped. At the same time perhaps the highest number of female-feticides, highest female-child mortality rate and highest number of dowry-related female deaths. In India this existence of two categories of females is very obvious. One is of respected, worshipped, and powerful females; and the other of totally unwanted, downgraded ones, who practically have no human rights. To look into its genesis, we have to peep back into the historical perspectives.

Historical Background

Indus Valley civilization existed prior to 3000 BC. This civilization and the other local tribal societies of that time, were all basically agrarian societies. The primitive agrarian societies, not only in India but all over the world, have had mother-right pattern. The reason is that, the agriculture was discovered by females, and it remained their exclusive occupation in its early stages. Females were believed to possess magical powers and procedures to have mastered the art of cultivation. For men, it was a pure magic. A peep into the agricultural rituals all over the world clearly indicates that the agricultural production was believed to be related to and vitally dependent upon the feminine powers, their reproductive functions, fertility and the other rituals of women. This resulted in economical and social supremacy of the females in the primitive agrarian societies, and the trebles having mostly matriarchal systems.

The social supremacy gradually shifted to the males among those societies that adopted the pastoral economy, as it depended more upon male's physical endowments. In those who continued with agriculture, female supremacy became more prominent. But, this female supremacy was gradually reduced with the introduction of cattle drawn plough. However, whenever there were any crop failures, droughts or insect attacks; the magical powers of the females, through ritualistic ceremonies, were requisitioned (Chattopadhyaya, 1973). The ritualistic traditions, signifying magical virtues of the female to communicate fertility to the field, are still seen in almost all countries of the world. Thus there was a basic difference between the two primitive societies; the one with agricultural base – had mostly matriarchal system with higher female status; and the other with the pastoral base – had male supremacy and primarily patriarchal system.

The influx of Indo-European Aryans, from Caspian Sea areas into the Indian sub-continent began around 1500 BC. Aryans had pastoral economy, followed patriarchal system, but their females had high social status. This higher female status is clear from the female deities and Goddesses worshipped by them. At that time, the Indus Valley Civilization had mostly disintegrated and the local population consisted primarily of agrarian societies or trebles with usually matriarchal system. As the Aryans settled down in India, two things could not be avoided in the normal course of events. The mixing up of the native population with Aryans through conjugal rights; and the socio-cultural impact of the native system of female supremacy on the Aryans' male-supreme patriarchal system.

In order to ensure their unchallenged and complete supremacy over the locals, the Aryans initiated a series of social and religious measures that ensured a strict hierarchy in the society to prevent the locals from mixing-up with the ruling class. One such measure was the introduction of the "Caste System" which, while dividing the society also lowered the female's social status. The caste system, known as *Varna* system, initially divided the society into two categories, i.e. rulers and workers. Gradually, it became more and more rigid and four distinct categories emerged. These were *Brahmins* (the priests who were the ideologues for the ruling and trading classes), *Kashytras* (the fighters and the rulers), *Vaish* (the traders), and the fourth category was of „*Shudras* " – the workers who ultimately became the untouchables. It was a gradual process which was initiated during the early *Vedic* period, i.e. ~1500–1000 BC and became more pronounced during the late *Vedic* period, i.e. around ~600 BC.

The early *Vedic* period (~1500–1000 BC), is also marked with the initiation of various types of rituals and religious ceremonies (*Yagnas*). The late *Vedic* period (~1000–600 BC), saw the emergence of several new Gods and the predominance of the *Brahmins*. During this period all religious and social ceremonies became so elaborate and complicated that to perform these, specially trained persons i.e., *Brahmins*, became indispensable. Under this *Brahminic* cult, sacrifices, rituals, and caste segregation in the society became extremely strong. All the working classes, which primarily consisted of the locals, were reduced to untouchables. The females and the untouchables were considered equivalent in most of the ritualistic ceremonies. This equivalency broke down the higher status of females amongst the locals. It became possible because with the loss of agricultural lands to Aryans, the productive usefulness of the ceremonies concentrated around the female sex organs was lost to the trebles and these became more ritualistic and distorted. Ultimately, it lead to the *Tantrik* forms of meditation and further down-gradation of

Table 1. The "5-M" ingredients of Tantric meditation

Mudra	Female
Matsya	Fish
Mamsa	Meat
Madira	Liquor
Maithuna	Sexual intercourse

the females to the level of only sex symbols (Table 1). The inferior status of the females was religiously and culturally propagated for centuries and it became the socially acceptable norm in India.

The emergence of two new religions around 600 BC, was the first socio-cultural uprising towards the complexities and ritualism of the Aryans and their Brahminic cult. These were the Jainism and the Budhism. Both these religions advocated non-violence and had no faith in rituals, sacrifices, religious ceremonies (Yagnas), or the caste system. Obviously both of these religions flourished at the cost of Vedic religion (the religion based on Vedas – the religious books of Aryan). Both of these religions being non-violent, and having had their base in the working classes were pushed into background, in India, during the first stage of revivalism of Vedic religion which started around 200 BC to 200 AD. It was during this period the famous scripts like "Manusamriti" and "Bhagwadgita" were written.

Manusamriti is the oldest living law book and is the foundation of Hindu law. It is believed to represent the true sense of Vedas. Manusamriti provided the legal sanctity to strict caste hierarchy and female subordination. It placed numerous restrictions on the females, and their sphere of action was limited inside the home only. Only those females, who begot sons, were put in better category. The sole purpose of female existence was legally limited to giving birth to males and serving the males (Table 2). Even then, the females belonging to higher castes were given better status and treatment than males of lower classes (Das 1962, Chattopadhyaya 1973).

Whole of the Vedic literature is in tough Sanskrit language. The working class and women were prohibited from even trying to read it. All the religious ceremonies were made extremely complicated, but these had to be performed, as through centuries, it had been culturally impregnated into the minds of people that unless they undertake such rituals they and/or their dead parents or sibs would not achieve salvation. So complete dependency on the Brahmins was firmly established. They dictated what they wanted. Bhagwadgita, the most revered and still popular medieval Hindu scripture, also equated women with lowly born (Table 3). As the caste distribution in a society is more rigidly defined and implemented, it ensures to the presence of distinct classes with distinct work classification. The

Table 2. Manusamriti (~200 BC – 400 AD)

Marital code
A wife may be superseded by another wife if she is bearing daughters only.

Niyoga ceremony
Appointment of a wife or a widow to procreate a son from the sexual union with an appointed male. (A girl born out of such union is illegitimate).

Daiva marriage
Gifting of daughter bedecked with ornaments, to a priest who officiates at a sacrifice ceremony.

Female creation
When creating females, the Creator allotted to women a love of bed, of their dishonesty, malice, and bad conduct.

Female rights
Nothing must be done independently by a woman even in her own house.

Table 3. Bhagwad Gita (~200 BC – 200 AD)

"For those who take refuge in Me, O Partha (Arjuna), though they are Lowly born, women, Vaisyas, as well as Shudras, they also attain the highest goal."

people, because of being born in a particular caste, are supposed and destined to follow the norms attributed for that class. If such a situation becomes acceptable, the distinct hierarchical set-up suiting to the ruling continues to work smoothly. Along with this class categorization, the simultaneous classification of female is linked in each caste and also in general. When one system of class classification in a society is strictly implemented the other one automatically gets enforced. This could be done either through force or in a better way by making cultural and social customs so strong that one could not easily think otherwise. Therefore, as casteism advanced in India, so did the female sub-servility of the male.

The *Vedic* cult got another impetus during the *Gupta* dynasty, which may be called as second revivalism, around 300–600 AD. During this period the *Jainism* and *Budhism* were more or less completely routed. This period, in fact, saw an emergence of the feudal set-up. This period also saw a very strong entrenchment of the temple nexus and revival of a highly ritualistic society. Numerous temples were built which played extremely important role in the continuation of the hegemonic culture and feudal set-up. Strict caste system and female subordination was advocated and severest punishments were inflicted on defaulters. However, all these strict customs and obeisance to complicated ritualistic formalities were more essential and strictly enforced for the lower classes and the females. The rigidity of caste based categories and with it the closely associated female down-gradation, was at its peak around 700 AD when "*Sati*" doctrine became the norm. In this ceremony, a widow is supposed to burn herself alive with her dead husband to achieve salvation. Probably this was a better recluse for the widows, because the religion-backed social castigation that had been imposed on the widows was much more cruel, humiliating, disgusting and torturous to live with (Table 4).

Thus, India had a society where, females were given higher social status – provided they served their husbands well and gave birth to sons. They were worshipped if they died as Sati. It was a very subtle and clever system that ensured the availability of dedicated females for serving the males. To serve the husband, to give birth to his sons and to look after his sons was the sole objective for the life of a female – which was religiously, socially and culturally accepted. It ensured that generation after generation and century after century, the females would continue to serve the males with utmost sincerity and while doing so also feel privileged. All this for achieving the salvation and to rid of the sins which they did during their previous birth!

The second cultural revolt against the Vedic values, came with the emergence of "*Nath*" and "*Sidh*" philosophy during 1000 and 1300 AD and later during the "*Nirguna Bhakti*" movement during 1200 to 1500 AD. All of these had base among the working classes. *Nath* and *Sidh* philosophy preached against the "*Tantrik*" traditions of meditations and also the sexual exploitation of the females. The *Nirguna Bhakti* movement also denounced the caste system and provided an exalted position to the females (Table 5). Another attempt on the revivalism of Vedic values,

Table 4. Status of widow (~300–600 AD)

- Inauspicious
- Debarred from ceremonial festivities
- Absolute self negation
- Celibate life
- No right to property
- No honey, meat, fish or betel leaf

Table 5. Nirguna Bhakti Movement (~1300 – 1500 AD)

- Origin in craftsmen, peasants, low castes
- Revolt against *Varna* system
- Exalted position to females
- Adopted female symbols for meditation

Table 6. Ramacharitmanas (Tulsi Das, ~1600 AD)

"Drum, uncultured, lowly born (Shudras), animals, and women need to be thrashed."

where the female position is downgraded, is obvious from the works of Tulsi Das (~1600 AD) who compared women with lowly born and animals (Singh, 1997; Table 6).

India Today

The commercial use of prenatal diagnostic techniques for sex-determination, followed by selective female feticide, was introduced in India around 1978, and it spread very quickly. In order to curb the increasing menace of female feticide, the Indian Government on 28th July 1994 enacted "The Pre-natal Diagnostic Techniques (Regulation and Prevention of Misuse) Bill, 1994". Under this Act, any private or Government institution after 1st January 1996 can not undertake any genetic test, unless the State Government specifically approves it for this purpose. We carried out a survey in 1997 on 378 individuals to ascertain the impact of this legislation and the attitude of present generation towards various aspects of prenatal sex determination and the associated female feticide. Earlier, we had carried out a similar survey on 200 persons in 1987. The trend seen from the two surveys is interesting. The desire for a male child remains. However, the data of two surveys indicate some marked differences. Compared to 1987, in 1997 there was reduction in the frequency of people who liked to go in for the abortion of fetus of unwanted sex (50% → 12%); or those favoring sex pre-selection (67% → 23%). The number of people who felt that prenatal sex determination should be legally permitted was also significantly reduced (73% → 33%) and so was the number of people who thought that permitting prenatal sex-determination would lead to population control (70% → 46%).

The question is do these figures indicate in reality a change in the basic attitude of the Indian society or at least that in north-west India from where this data has been collected? The figures obtained do indicate a healthy change of perception, but unfortunately, I fear, these figures may not represent an actual change in the perception of the Indian society as such towards the females. The data collected is from individuals belonging to upper middle class families. These individuals are usually better educated, have smaller families, adopt more family planning measures, and have better living standards. But, they are also better informed about the prevailing laws. Therefore, the biasness in their responses can not be completely ruled out, since they knew that sex-determination was now legally prohibited. In any case, this indication of healthy change, unfortunately, is limited to an extremely small proportion of population and the male preference continues unabated in other sections of the society.

The preference for a male child is neither linked to literacy nor to economic status. For example Chandigarh, which is capital of two State Governments and one Union Territory, enjoys the highest educational and economic standards. But, in this city, the number of females per thousand males, at birth is amongst the lowest in India (1996: 771 females to 1000 males). The data regarding sex ratio at birth, obtained from Amritsar (1996: 770 females to 1000 males), is another example of unchanged attitudes. For all practical purposes the Prenatal Act of 1994 has not made any impact. The only difference, which this Act has made, is that rampant public hoarding and advertisements in the newspapers are not being given. These are now being given in concealed but obvious forms. The test reports are not given in black and white but conveyed orally. More and more centers are continuing to mushroom in all the cities and they would soon be invading villages. It is for the State Governments to implement the Pre-natal Diagnostic Act. However, most of the State Governments have not been pursuing it seriously enough and the tests go on.

One of the reasons contributing to the perpetuation of female child's unwanted status in India, is the new wave of revivalism for *Vedic* values which is going on at the present moment, utilizing mass media and satellite technologies. A closer look into the viewer's ratings of the Indian television channels reveals that the most popular serials are the semi-religious epics of the late *Vedic* periods (i.e. ~1000–600 BC) and those of *Gupta* dynasty (~300–600 AD). These epics distinctly depict superiority of the ruling class over the working class, superiority of the males, and also the religiously and socially justified sub-servility of the females. Many of these also do distinctly depict the females as symbols of worship. But which females are to be worshipped? These are again those who had dedicated their whole lives serving the males – i.e., either their husbands or their sons!

Historically in India, unlike Europe, the development had been uneven because those from lower castes and trebles were not allowed to join in the main stream. All the political or economic changes and socio-religious reforms remained limited to the top few in society. For the remaining millions, there was no change. For them, the cultural continuity remains unaffected. They still perform the same rituals, nurture the same beliefs, and hold the same attitudes towards the females – as it was centuries ago. These beliefs are so deeply permeated and pervasively imbedded in the minds of most Indians, that their perennially continuing influence can not be easily countered.

The increasing corruption in India, and the degradation of moral values are continuing to strengthen the medieval period attitudes where male was supreme. In spite of westernization of the urban Indian society, there has not been much advancement towards the cultural transformation, which remains at the level of the medieval period. The incidences of dowry-related deaths, maltreatment, bride burning, rapes, etc., have escalated. When females are at the receiving end, the parents in their hearts pray that they should not get a female child. The preference for a male child is a cultural and social reality of India. The mode of begetting a male child or not getting a female child is what differs. Those who are less religious go in for the abortion of female fetuses and those who are more religious or do not want to kill the fetus of unwanted sex (obviously female) go in for "sure-male" concoctions, quackery medications, *tantric* rituals, prayers, etc.

It appears that in India, the females are going to remain at the receiving end for more time. Legislation is essential to control any evil, but the mere enactment of Law would not remove the evil of female-feticide. Unless the people are totally involved in the eradication of this evil, little can be achieved. It is essential to apprise the public about the ills of this more or less socially acceptable custom. The level of public consciousness needs to be raised by educating the common people about the legal sanctions against this custom. Concerted efforts would have to be made to break the existing caste system, bring in cultural consciousness highlighting the importance of females as equal partners, and to stop the medievalization of the present generation. The great cultural efforts which, are needed to break the present male-preferred attitude of the society – are unfortunately not visible in India.

References

Chattopadhyaya D (1973) Lokayata. A study in ancient Indian materialism. People's Publishing House, New Delhi, India

Das RM (1962) Women in Manu and his seven commentators. Kanchana Publications, Bodh-Gaya, India

Singh S (1997) Kabir's view on female (In Hindi) Nirmal Publications, Delhi, India

Session III

Can the Implementation of New Genetic Tests Be Safeguarded by Consensus Policy Recommendations?

Ethics and Genetics in International Perspective: Results of a Survey

Dorothy C. Wertz

Full discussion of the ethical provision of genetic services can take place only with adequate knowledge of the views and practices of those involved.

In order to find out how the various stakeholders viewed ethical issues that occur in medical genetics, we surveyed 4621 genetics services providers in 37 nations, including 1538 in the US, 846 primary care physicians (pediatricians, obstetricians, and family practitioners) in the United States, 718 first-time visitors to 12 genetics clinics in the United States and Canada, and 1000 members of the general public. Patients received the questionnaires before their first clinic visit, in order to insure that their ethical views were not influenced by genetic counseling. Members of the public were surveyed by printed questionnaires delivered in person, using a quota sample developed by Roper-Starch Worldwide, a professional survey firm.

In all, 2902 (63%) geneticists around the world returned questionnaires, including 1084 (70%) of those who received questionnaires in the United States and 102 (47%) in the United Kingdom. Response rates appear in Table 1. Professional characteristics of European and US respondents appear in Table 2. 499 (59%) primary care physicians, 476 (66%) patients, and 988 (99%) members of the public responded. Most patients lived in the United States (93%), were women (91%) and white (89%). They had a median of 13 years of education; 13% had not finished high school, and 28% were college graduates (including 11% with some postgraduate training). Most were in working class occupations (clerical or sales, 29%, service, 19%, or factory production work, 8%, with a minority in professional or managerial work, 30%). The majority of their spouses were also working class, with 31% in factory production work, 23% in clerical or sales, and 15% in service occupations. Median income was US $25,000–45,000. This group appeared closer to „middle America" than the college-educated respondents or members of consumer support groups sometimes surveyed about „genetic discrimination" (Lapham et al., 1996).

In the following tables we report responses of the 409 who were parents of minor children. The median age of child patients was five years, and the most fre-

Table1. Survey response rates

Country	Invited to participate	Responded	% Responding
Argentina	57	19	33
Australia	26	15	58
Belgium	40	15	38
Brazil	131	74	56
Canada	212	136	64
Chile	25	16	64
China	392	252	64
Colombia	27	14	52
Cuba	96	14	16
Czech Republic	137	81	59
Denmark	54	28	52
Egypt*	2	2	100
Finland	53	22	42
France	102	75	74
Germany	418	255	61
Greece	12	12	100
Hungary	78	36	46
India	70	23	33
Israel	27	23	85
Italy	23	21	91
Japan	174	113	65
Mexico	89	64	72
Netherlands	41	27	66
Norway	18	9	50
Peru	16	14	88
Poland	250	151	60
Portugal	22	11	50
Russia	66	46	69
South Africa	21	16	76
Spain	82	51	62
Sweden	15	12	80
Switzerland	10	6	60
Thailand	28	25	89
Turkey	30	22	73
U.K.	217	102	47
U.S.A.	1538	1084	70
Venezuela	22	16	73
Total	4621	2902	63

* Removed from data analysis because of low number asked to respond.

quently mentioned conditions were neurofibromatosis, Marfan syndrome, Down syndrome, cleft lip/palate, and cancer syndromes.

Countries were chosen so as to include all nations with a minimum of ten practicing medical geneticists as of 1993. Large areas of the world, such as sub-Saharan Africa, have few geneticists. In each country, a geneticist colleague drew up a list of persons practicing genetics, and distributed and collected the questionnaires. In

Table 2. Professional and personal characteristics of respondents

Professional characteristics	U.S.	35 other nations
Degree		
MD	32%	72%
(Pediatrician)	24%	31%
PhD	18%	13%
MS (genetic counselor)	34%	5%
Other	16%	10%
Years in genetics (median)	9	9
Patients per week (median)	6	1-5

Personal characteristics		
Median age	40	42
Women	72%	55%
Experience with disability outside professional role	77%	62%
Religious background		
None	20%	34%
Catholic	18%	32%
Protestant	34%	18%
Jewish	22%	3%
Other	6%	13%
Attendance at religious services		
Less than once a year	30%	47%
several times a year	35%	32%
1-3 times a month	16%	8%
Weekly or more	19%	13%

the United States, Canada, and United Kingdom, specially-trained genetic counselors and genetic nurses without doctoral degrees were included, because they do much of the actual counseling. In Poland, our colleague included midwives, who do much of the prenatal counseling.

Questionnaires were sent out in three waves of mailings, followed in some countries by a telephone reminder. All questions were answered anonymously.

The questionnaires included socio-demographic data and 50 questions on ethical situations, mostly presented in the form of case vignettes. Respondents were asked to select their preferred course of action in each case, from a checklist, and to write, in their own words, why they had chosen this course of action. At the end of the questionnaire, they were asked to list the three questions they found most difficult to answer.

The „most difficult" questions appear in Table 3. In what follows, I will concentrate on 1) counseling after prenatal diagnosis; 2) disclosure to relatives; 3) disclosure to spouse, including nonpaternity; 4) presymptomatic testing of children; 5)

Table 3. Most difficult questions

Questions	%Listing
Which disorders to abort	41
Parenthood for persons with disabilities	28
Sex selection	29
Disclosure to relatives	25
Huntington disease questions	17
Disclosure to spouses	24
Third party to genetic information	17

views on autonomy; 6) sex selection; and 7) third party access to genetic information. Tables will show data from Europe and the United States.

Counseling after Prenatal Diagnosis

„Nondirective counseling" is frequently presented as the ideal (Fraser 1974). In our earlier survey of 19 nations in 1984–85 (Wertz and Fletcher 1989), close to 100% of respondents said they subscribed to the various different goals and practices that constitute nondirectiveness. Yet when presented with an actual situation, a fetus diagnosed with one of 24 different conditions, the majority of respondents to our 1993–95 survey would be directive, either through openly telling patients what to do or through presenting purposely slanted information as scientific/medical fact. „Nondirectiveness," trying to be „as unbiased as possible", appears to be an aberration from the world norm. Nondirectiveness prevailed mostly in the United States, Canada, United Kingdom, and Australia. On the European continent, Norway, the Netherlands, Germany and Switzerland were the least directive, while Belgium, Greece, France and nations in Eastern Europe were the most directive. Tables 4, 5, and 6 show the percents who would be nondirective about fetuses with various conditions appearing in childhood and adulthood. Respondents had five choices: urge parents to carry to term; emphasize positive aspects so they will favor carrying to term without suggesting it directly; try to be as unbiased as possible [nondirectiveness]; emphasize negative aspects so they will favor termination without suggesting it directly; urge termination. (A sixth choice, not tell them this particular test result, found little favor in Western nations.) Pessimistic counseling (emphasize negative aspects or urge termination) outweighed optimistic counseling for severe, open spina bifida, trisomy 21, cystic fibrosis, sickle cell anemia, achondroplasia, neurofibromatosis, familial hypercholesterolemia, and Huntington disease. Optimistic counseling prevailed (though often by only a few percentage points) for XXY (Klinefelter syndrome), 45,X (Turner syndrome), PKU in the fetus, severe, untreatable obesity in the absence of a known ge-

Table 4. Counseling after prenatal diagnosis for conditions appearing in childhood

Country	% would provide unbiased counseling for				
	Severe spina bifida	Trisomy 21	Cystic Fibrosis	Achondro-plasia	Sickle cell Anemia
Northern/Western					
Europe	42	59	61	64	65
Belgium	23	33	39	79	36
Denmark	32	54	58	69	68
Finland	52	71	71	81	71
France	15	28	32	43	50
Germany	49	68	70	69	69
Netherlands	46	72	69	70	80
Norway	67	78	67	89	78
Sweden	33	42	42	55	42
Switzerland	67	67	67	83	67
United Kingdom	63	84	87	87	88
Southern Europe	34	36	42	32	43
Greece	9	9	9	9	27
Italy	41	48	59	57	59
Portugal	36	46	55	46	55
Spain	36	35	38	23	36
Eastern Europe	17	27	30	36	39
Czech Republic	7	5	12	23	31
Hungary	3	19	11	26	33
Poland	27	47	45	51	45
Russia	12	9	29	21	41
U.S.A.	70	86	88	88	88

netic syndrome, and predisposition to bipolar disorder, schizophrenia, alcoholism, or Alzheimer's disease (assuming that these predispositions could be tested for).

In Europe, most of the „directiveness" was through provision of purposely slanted information rather than by openly telling people what to do. Presenting slanted information as „fact" is ethically worse than open directiveness, because usually the patient has no reason to suspect that the information is biased and has insufficient knowledge to question the expert's veracity. If the expert openly gives his/her opinion about what to do, the patient at least has the opportunity to disagree. Presenting slanted information under the guise of fact is the equivalent of propaganda, which is meant to mislead and confuse. Counseling was especially pessimistic in Eastern Europe, Greece, and Belgium.

In a related set of questions, most geneticists took a pessimistic view of disability in general. Patients and primary care physicians shared this view. Few in any group had experience with support groups.

Table 5. Counseling after prenatal diagnosis for conditions appearing in childhood

Country	% would provide unbiased counseling for				
	Neurofi-bromatosis	XXY	45,X	PKU in fetus	Severe Obesity
Northern/Western					
Europe	65	48	49	46	50
Belgium	77	54	62	54	39
Denmark	69	81	85	58	69
Finland	67	62	67	67	52
France	56	34	39	23	20
Germany	64	42	41	46	56
Netherlands	72	67	67	67	57
Norway	78	100	100	56	67
Sweden	58	67	67	42	55
Switzerland	83	33	50	50	33
United Kingdom	80	72	73	72	65
Southern Europe	44	36	35	36	36
Greece	27	36	18	9	27
Italy	67	55	59	50	55
Portugal	73	73	82	55	55
Spain	32	19	17	32	26
Eastern Europe	42	39	42	32	42
Czech Republic	31	36	44	30	51
Hungary	28	33	39	17	53
Poland	53	45	43	39	30
Russia	39	31	34	21	59
U.S.A.	89	84	81	77	81

Disclosure to Relatives, Against a Patient's Wishes

Sometimes a patient refuses to disclose information that could be useful to blood relatives at genetic risk. This poses a dilemma between two well-known duties in medicine: the duty to preserve patient confidentiality and the duty to warn third parties of harm. In 1983, the US President's Commission recommended that confidentiality might be overridden in certain exceptional circumstances, provided that the following four conditions were met: „1) reasonable efforts to elicit voluntary consent to disclosure have failed; 2) there is a high probability both that harm will occur if the information is withheld and that the disclosed information will actually be used to avert harm; 3) the harm that identifiable individuals would suffer would be serious; and 4) appropriate precautions are taken to ensure that only the genetic information needed for diagnosis and/or treatment of the disease in question is disclosed" (p 44). In 1994, the Institute of Medicine reaffirmed this statement. In 1997, the American Society of Human Genetics (ASHG) approved a

Table 6. Counseling after prenatal diagnosis for adult-onset disorders

Country	% would provide unbiased counseling for				
hypercholes-	Familial disease terolemia	Huntington to mental	Predispos. to illness	Predispos. Obesity alcoholism	Predispos. to Alzheimer
Northern/Western Europe					
Belgium	64	57	77	64	64
Denmark	62	65	81	80	86
Finland	67	57	76	71	72
France	49	54	58	31	49
Germany	63	71	56	51	57
Netherlands	69	83	62	67	72
Norway	89	78	89	100	89
Sweden	50	58	75	92	68
Switzerland	67	80	67	33	50
United Kingdom	76	87	79	76	83
Southern Europe					
Greece	36	9	36	36	27
Italy	55	64	73	73	68
Portugal	91	64	82	73	73
Spain	33	48	43	35	35
Eastern Europe					
Czech Republic	39	30	54	48	44
Hungary	42	25	53	61	44
Poland	49	51	54	51	49
Russia	65	33	58	50	42
U.S.A.	83	86	86	83	85

„Points to Consider" statement on „Professional Disclosure of Familial Genetic Information" that generally follows the President's Commission guidelines, except that ASHG specifies that the disorder must be treatable. The ASHG statement does not discuss whether use of information for the relatives' reproductive planning is a legitimate cause for overriding confidentiality.

Dr. Fletcher and I have long argued that genetic information should be regarded as familial, rather than individual property. There was no consensus in North America and Northern Europe on this issue (Tables 7, 8, and 9). Roughly half would respect confidentiality. In Eastern Europe and Southern Europe, however, the majority would tell the relatives, especially if they asked. These responses suggest a different view of individual privacy from the views held in nations of Northern Europe that were strongly influenced by the 18th-century Enlightenment. Some respondents from Southern or Eastern Europe said that they did not consider it a breach of confidentiality to tell relatives who asked; only if the doctor

Table 7. Telling relatives at risk about a genetic diagnosis: International views. A man diagnosed with Huntington disease refuses to tell his siblings and children, who are at 50% risk.(n=2902 geneticists in 37 nations)

Geographical area	% would			
	Preserve confidentially	Tell relatives only if they ask	Tell relatives unasked	Refer to family physician
U.S.A.	53	32	6	9
Other English speaking nations	45	40	5	10
Northern and Western Europe	52	34	2	12
Southern Europe	26	39	14	21
Eastern Europe	22	42	15	21
Near East	26	51	13	10
Asia	27	29	31	12
Latin America	20	45	18	16

Table 8. Disclosure of a cancer mutation to relatives, against patient's wishes

Country	% would			
	Respect confidentially	Tell relatives if they ask	Tell relatives unasked	Let referring doctor decide
Northern/Western Europe	42	35	7	15
Belgium	29	50	14	7
Denmark	54	27	4	15
Finland	35	30	5	30
France	25	41	11	22
Germany	51	35	3	12
Netherlands	19	23	35	19
Norway	33	44	11	11
Sweden	58	25	8	8
Switzerland	25	50	0	25
United Kingdom	34	44	9	13
Southern Europe	20	31	22	27
Greece	20	10	40	30
Italy	41	12	23	24
Portugal	9	18	27	46
Spain	15	46	17	23
Eastern Europe	20	33	23	24
Czech Republic	17	36	28	19
Hungary	17	41	0	41
Poland	23	34	17	26
Russia	16	21	48	16
U.S.A.	47	30	12	11

Table 9. A woman tests positive for a breast cancer gene

% would	Country				
	U.S.A.		Germany	Japan	Hungary
	Counselors	Geneticists			
Tell her sister she is at risk, against the patient's wishes	25	46	44	44	48
Encourage patient's 20 year-old daughter to be tested	54	75	48	42	52
Encourage 13 year-old daughter to be tested now	13	27	7	22	13
Suggest prophylactic double mastectomy for patient	11	18	4	0	4
Have counseled for breast cancer genes	36	37	13	1	13

went out and found the relatives, and told them, unasked, would this be a breach of confidentiality. Sixty percent of US patients favored disclosure.

Disclosure to Spouse

Few geneticists in Northern Europe or North America thought a spouse or partner should have access to genetic information without an individual's consent, even for a translocation that might cause Down syndrome in the couple's children (Table 10). More would disclose the translocation in Southern Europe (39%) or Eastern Europe (44%). Interestingly, more patients (42%) than providers (20%) in the US favored disclosure to spouse.

Almost no one would tell a husband unasked about an „accidental finding" that he is not the father of a child. If he asks directly, a minority would tell him (36% in the US, 17% in Northern Europe, 32% in Southern Europe, and 15% in Eastern Europe). Geneticists in the United States said that this was none of their business, unless a test were done specifically for purposes of identifying paternity. The majority of patients (75%) – most of whom were women – thought the doctor ought to tell a man who asked, though 57% said the doctor ought to warn the woman first.

Table 10. Spouse/partner's access to information: International views

Geographical area	% thought spouse/partner should have access without consent					
	Translocation Down	Huntington disease	Cystic Fibrosis (CF)	CF carrier	Schizophrenia	Predisposition to Alcoholism
United States	20	9	6	3	4	4
Other English speaking nations	17	11	6	4	5	7
Northwest Europe	17	6	6	3	3	3
Southern Europe	39	32	27	23	24	19
Eastern Europe	44	30	24	20	25	26
Near East	40	26	21	11	23	23
Asia	39	37	34	24	31	33
Latin America	56	37	32	26	31	30

Testing Children and Adolescents for Adult-Onset Disorders

In the United States, large majorities of geneticists, primary care physicians, and parents favored testing children and adolescents for familial hypercholesterolemia and cancer syndromes (Table 11). About half of geneticists and three- quarters of primary care physicians would test children and adolescents for genetic susceptibility to alcoholism. (Many US physicians believe that alcoholism may be prevented if susceptible people are identified early, but there is little evidence for this). About one-quarter of geneticists would test for Huntington disease or Alzheimer's disease. Fewer women than men would test for these conditions (Wertz, 1997).

Respondents were asked to explain, in their own words, their responses regarding testing children and adolescents. We recorded two reasons behind each write-in comment, using categories developed in the 1984–85 survey (Wertz and Fletcher, 1989). In giving reasons for their responses, 66% said they wished to avoid harm, though early treatment or prevention for familial hypercholesterolemia, cancer, or possible alcoholism, or to avoid harm from premature knowledge of Huntington disease or Alzheimer's disease. 43% said the information was necessary to make decisions about medical management. Fewer mentioned parental autonomy (13%) or preparing for the future (9%) as reasons for testing. More (25%) mentioned protecting the child's autonomy as a reason not to test, and 13% said there was no need to know at this time. Few (3%) mentioned the possibility of stigmatization as a result of testing. A substantial minority (38%) had had requests to test children for adult-onset disorders. The most frequently reported condition for which parents sought testing was Huntington disease.

In contrast to geneticists, the majority of both primary care physicians and parents thought parents should be able to have their minor children tested for both Huntington and Alzheimer's disease. Most members of the US public thought parents should be able to have their children tested for conditions that were „treatable if found early" (84%), or „preventable" (81%). A slight majority (53%) approved testing for conditions that were „neither preventable nor treatable." These results point to a dichotomy between geneticists and the rest of the US medical community, whose views closely paralleled those of their patients.

In a separate question, asked only in the United States, Germany, Japan, and Hungary, 27% of US geneticists and 13% of US genetic counselors would urge a woman with a dominant breast cancer gene to have her 13-year-old daughter tested now (Table 9); 71% of parents said it was moderately (22%), very (17%), or extremely (32%) likely that if they themselves had a breast cancer gene they would encourage their own 13-year-old daughters to be tested now (not shown in table). Some parents wrote in that they would impose radical lifestyle changes on a daughter who had a breast cancer gene, including a mother who, apparently believing that pregnancy prevented breast cancer, said that she would make her

Table 11. Testing children: International perspectives (n=2902 providers)

Country	% who thought parents should be able to have minor children tested					Had requests to test children for adult-onset disorders	Would tell a minor results of Alzheimer test
	Huntington disease	Alzheimer disease	Alcoholism (susceptibility)	Cancer genes	Familial hyper-cholesterolemia		
English speak. nations (total)	9	9	19	62	69	65	16
Australia	13	13	7	64	73	93	0
Canada	12	11	26	63	71	56	22
South Africa	25	25	67	81	81	50	0
United Kingdom	2	2	4	58	67	76	12
United States	27	25	48	70	81	38	18
Northern/Western Europe (total)	16	14	27	51	63	49	12
Belgium	15	15	50	62	92	57	17
Denmark	33	33	33	56	67	37	14
Finland	18	18	23	41	64	41	13
France	21	19	37	72	85	51	9
Germany	14	10	24	45	56	44	12
Netherlands	11	11	18	36	57	84	13
Norway	11	11	11	56	78	67	0
Sweden	8	8	25	50	42	58	0
Switzerland	40	20	33	80	67	83	20
Southern Europe (total)	66	62	66	80	89	34	1
Greece	90	70	70	80	90	40	0
Italy	48	48	57	76	91	38	0
Portugal	64	64	64	82	91	27	0
Spain	70	66	70	82	88	33	2
Eastern Europe (total)	73	70	78	88	93	16	5
Czech Republic	73	70	78	86	95	30	5
Hungary	58	56	58	75	81	17	0
Russia	86	80	93	91	93	12	7

daughter get pregnant at age 16 and would keep her pregnant every year thereafter.

The low percents who would test in the United Kingdom are especially striking (Table 12). This reluctance to test may stem in part from greater experience with requests for testing children in the U.K. than anywhere else in the world except Switzerland and the Netherlands. In the U.K., 76% had had requests to test children for adult-onset disorders, as compared with 38% in the US and 49% in Western Europe. Percents in Northern/Western Europe who would test are also lower than in the US (Table 11). In Southern Europe, and Eastern Europe, however, majorities would test children for all conditions listed (Table 11). Willingness to test in these regions appears to stem from cultural beliefs about the rights of parents and their authority over their children.

European data appear in Table 11. Denmark and Switzerland stand out from other North/Western European nations as having higher percents who would test for Huntington or Alzheimer's disease, though they were still in the minority. Russia, the Czech Republic, and Greece had the highest percents who would test for these disorders in Southern/Eastern Europe, while Italy had the lowest percents. As in the Untied States, fewer women than men geneticists would test for these conditions.

The Clinical Genetics Society (U.K.) (1993) American Medical Association (1995), and American Society of Human Genetics (1995) have urged caution about testing of children.

The Trend Toward Increased Autonomy

Americans' overwhelming cultural belief in autonomy was reflected in survey responses. For example, 4% of geneticists in the U.K. and 36% in the US thought patients were entitled to whatever services they could pay for out-of-pocket; 93% in the U.K. and 90% in the US thought professionals owed patients a referral for procedures that the professional morally opposed; 43% in the U.K. and 65% in the US thought they owed referrals out-of-state or out-of-country; 14% in the U.K. and 55% in the US thought they owed referrals for sex selection, 46% in the U.K. and 90% in the US thought parents had a right to know fetal sex.

Many physicians appear to have forgotten that they can ethically refuse a service that provides no known medical benefit. Instead, they feel that they have to honor all patient requests, or at least offer a referral. Fear of lawsuits did not emerge as an issue in the write-in comments. Although much US medical practice is driven by lawsuits, patients cannot sue for refusal of services that are not customary, including sex selection or testing minors for adult-onset disorders or carrier status.

Increased respect for patient autonomy appears to be a belief spreading to other parts of the world, as indicated by some comparisons between responses to our 1984–85 survey and the 1993–95 survey. For example, the majority in the recent survey would tell a woman about her XY status, as compared to 50% in the earlier survey.

Readers should pay special attention to Line 7 (parents' views) in Table 12. In general, parents were more autonomy-oriented than either group of providers. (The major exception, parents' denial of a „right not to know," resulted from their belief, expressed in written comments, that no one would take a test in the first place unless they wanted to know). The parents' views may represent the wave of the future, particularly in the United States, as medicine becomes a set of big businesses that try to market themselves to consumers. What consumers want is every service they ask for, without limit. They believe that nothing should be withheld (this would be a „denial of patients' rights") and that patients are „entitled" to whatever service they request, as long as they can pay for it out-of-pocket.

Sex Selection

The trend toward respect for autonomy is evident in responses to the questions about sex selection. In all but three of the 19 nations surveyed in 1984–85 (India, France, and Sweden), more geneticists would perform prenatal diagnosis for sex selection (in the absence of an X-linked disorder) than would have done so in the earlier survey (Table 13). This increase in willingness to accede to requests for sex selection has occurred in spite of much ethical discussion in the intervening years condemning sex selection. Most respondents who would provide the service gave respect for autonomy as their major reason (Wertz and Fletcher 1998). They may also have believed that sex selection was not a social problem in Western nations, because most people prefer a balanced family that includes children of both sexes. Many said that children had a right to be born into families where they were wanted. Those who would refuse sex selection said they opposed abortion of a normal fetus or that sex selection was a „slippery slope" toward various cosmetic selections unrelated to health. Few in Europe or North America mentioned the position of women or a need to maintain a balanced sex ratio, though this type of reasoning was relatively common in India. (One reason why fewer in India would do sex selection than in the earlier survey may be that India has made sex selection illegal.)

Table 12. Views on patient autonomy (n=2902 geneticists)

Country	% agreeing				
	Withholding any requested service is paternalistic	Patients are entitled to any service they request and can pay for out-of-pocket	After taking a test, patients have right not to know results	I should offer referral if unwilling to perform a procedure for moral reasons	Counselor should support patients' decisions even if disagrees
English speaking nations					
Australia	50	27	87	100	87
Canada	38	11	90	94	86
South Africa	56	38	75	88	100
United Kingdom	51	4	88	93	99
United States (n=1084)	56	36	82	90	90
U.S. Primary care physicians (n=499)	57	26	62	82	n/a
U.S. Patients (n=409)	69	59	41	86	79
Northern/Western Europe					
Belgium	53	0	80	100	80
Denmark	67	19	93	87	74
Finland	29	0	87	91	91
France	45	8	78	100	49
Germany	28	13	89	81	74
Netherlands	37	7	96	93	93
Norway	56	22	78	78	100
Sweden	58	8	75	92	83
Switzerland	67	33	100	83	100
Southern Europe					
Greece	50	33	50	83	83
Italy	55	35	81	95	81
Portugal	36	27	100	100	91
Spain	41	20	50	88	65
Eastern Europe					
Czech Republic	48	41	63	79	45
Hungary	15	29	86	80	63
Poland	67	68	62	91	54
Russia	55	67	37	81	41

Table 13. Responses to outright requests for sex selection (n=2902 geneticists)

Area/country	% would perform prenatal diagnosis for sex selection and provide requested information (in parentheses: additional % who would refer)				
	1. Single woman wants girl	2. Couple with 4 girls wants boy	3. Poor couple with 5 boys wants girl	4. Nonwestern couple wants boy	5. Couple in 40's wants girl
English speaking nations					
Australia	21 (7)	21(29)	29(36)	29(29)	36(21)
Canada	17(33)	17(34)	21(32)	21(39)	38(29)
South Africa	14(14)	21(14)	20(20)	21(29)	29(14)
United Kingdom	8(25)	12(27)	14(24)	17(34)	22(27)
United States	35(36)	34(38)	38(37)	38(38)	57(28)
[United States (public)]	35	38	41	–	–
Western Europe					
Belgium	36 (7)	36 (7)	20(27)	43(21)	53 (7)
Denmark	22 (4)	22 (4)	27	26 (7)	46 (4)
Finland	14 (5)	19 (5)	32(14)	33(14)	57(10)
France	4 (3)	8 (1)	9 (1)	10(13)	24 (4)
Germany	11 (8)	13 (7)	14(11)	25(11)	23 (7)
Greece	25 (8)	33 (8)	30	50(17)	50 (8)
Italy	29(10)	25(10)	27(18)	30(15)	35(10)
Netherlands	0(23)	0(23)	0(35)	12(31)	16(24)
Norway	13	13	11	13	25
Portugal	46(18)	36(46)	46(27)	36(46)	55(36)
Spain	23 (2)	23 (2)	15 (2)	27 (2)	42 (2)
Sweden	11(11)	11(11)	25 (8)	22(11)	22(11)
Switzerland	0	0	0	0(33)	17
Eastern Europe					
Czech Republic	37 (5)	49 (4)	54(10)	56(12)	68 (4)
Hungary	26 (9)	63(14)	78 (3)	64 (3)	61(17)
Poland	29 (7)	30(16)	28 (4)	48(12)	46 (9)
Russia	78 (3)	90 (5)	72 (9)	85 (7)	91 (7)

Table 13. *Continue*

Area/country	% would perform prenatal diagnosis for sex selection and provide requested information (in parentheses: additional % who would refer)				
	1. Single woman wants girl	2. Couple with 4 girls wants boy	3. Poor couple with 5 boys wants girl	4. Nonwestern couple wants boy	5. Couple in 40's wants girl
Near East					
Egypt	0	0	0	0	0
Israel	67(10)	68(14)	70 (5)	82(14)	82 (5)
Turkey	0	10	14 (9)	5(11)	18
Asia					
China	34 (6)	24 (3)	28 (2)	35(14)	29 (4)
India	27	32	19	32(9)	46
Japan	10 (2)	18(2)	22 (2)	19(11)	19 (3)
Thailand	8	8	4	8 (4)	16
Latin America					
Argentina	33	25	31 (8)	25	55
Brazil	35(10)	34(11)	37 (7)	32(16)	53(10)
Chile	7 (7)	13 (7)	13	7	25
Colombia	25(17)	25(17)	23(15)	50(17)	67 (8)
Cuba	46	62	57 (7)	54 (8)	85
Mexico	39 (3)	38 (3)	28 (2)	39 (5)	53 (3)
Peru	39	39	36 (7)	57	50
Total	27(20)	29(20)	31(20)	35(24)	44(15)
Total excluding U.S.	23 (9)	26 (9)	27(10)	33(15)	39 (9)

* See table 4 for full case descriptions

Table 14. Sex selection: Trends in geneticists' willingness to perform, 1985-1994
Case: A couple with 4 healthy daughters desire a son. They ask for prenatal diagnosis (PND) to find out the fetus's sex. If it is a girl, they will terminate the pregnancy.

Area/country	% would					
	Perform PND		Offer a referral		Total would perform or refer	
	1985	1994	1985	1994	1985	1994
English speaking nations						
Australia	9	21	8	29	17	50
Canada	30	17	17	34	47	51
South Africa	–	21	–	14	–	35
United Kingdom	9	12	15	27	24	39
United States	34	34	28	38	62	72
Western Europe						
Belgium	–	36	–	7	–	43
Denmark	13	22	0	4	13	26
Finland	–	19	–	5	–	24
France	7	8	6	1	13	9
Germany	6	13	1	7	7	20
Greece	29	33	0	8	29	41
Italy	18	25	0	10	18	35
Netherlands	–	0	–	23	–	23
Norway	17	13	0	0	17	13
Portugal	–	36	–	46	–	82
Spain	–	23	–	2	–	25
Sweden	28	11	10	11	38	22
Switzerland	0	0	0	0	0	0
Eastern Europe						
Czech Republic	–	49	–	4	–	53
Hungary	60	63	0	14	60	77
Poland	–	30	–	6	–	36
Russia	–	90	–	5	–	95
Near East						
Egypt	–	0	–	0	–	0
Israel	13	68	20	14	33	82
Turkey	0	10	20	0	20	10
Asia						
China	–	24	–	3	–	27
India	37	32	15	0	52	32
Japan	6	18	0	2	6	20
Thailand	–	8	–	0	–	8

Third Party Access to Genetic Information

Almost nobody (fewer than 1%) thought that employers, life insurers, or schools should have access to genetic information without a person's consent. The world-wide distrust of institutional third parties, especially insurance companies, was such that in many nations almost half thought that these entities should have no access at all to genetic information, even with the individual's consent.

Reports of patients being refused employment or insurance on the basis of carrier status, presymptomatic testing, or genetic susceptibility were virtually nonexistent outside the United States. Many respondents said that in their countries laws prevented such things from occurring. Even in the United States, reports of refusals (whether from geneticists, primary care physicians, or patients) were few in comparison with the volume of patients seen, and appeared to be part of the vagaries of insurance practice generally, rather than examples of „genetic discrimination."

Summary

Overall, the survey results suggest a need to 1) re-examine the process of genetic counseling, with a view to presenting more balanced information; 2) re-consider whether the „patient" in genetics might include the genetic family as well as the individual; 3) recognize that patients may see an ethical problem very differently than do providers; 4) consider whether there should be limits to autonomy, or whether any attempt to set limits might do more harm than good; 5) provide geneticists with experience with persons with disabilities, outside the geneticists' professional role, perhaps through liaisons with support groups.

References

American Medical Association Council on Ethical and Judicial Affairs (1995) Testing Children for genetic status. Code of Medical Ethics, Report 66. American Medical Association, Chicago

American Society of Human Genetics Board of Directors and American College of Medical Genetics Board of Directors (1995) Points to consider: ethical, legal, and psychosocial implications of genetic testing in children and adolescents. Am J Hum Genet 57:1233

American Society of Human Genetics, Social Issues Committee (1998) Points to Consider; Professional Disclosure of Familial Genetic Information. Am J Hum Genet in press

Clinical Genetics Society (1994) The genetic testing of children: report of a working party of the Clinical Genetics Society, Angus Clarke, Chair. J Med Genet 31:785–797

Fraser FC (1974) Genetic counseling. Am J Hum Genet 26:636–659

Institute of Medicine, Committee on Assessing Genetic Risks (1994) Assessing Genetic Risks. National Academy Press, Washington, D.C., p 276

Lapham EV et al (1996) Genetic discrimination: perspectives of consumers. Science 274:621

United States, President's Commission for the Study of Ethical Problems in Medicine and Biomedical and Behavioral Research (1983) Screening and Counseling for Genetic Conditions.
Washington, DC: US Government Printing Office, p 44

Wertz DC (1997) Is there a „women's ethic" in genetics?: a 37-nation survey of providers. Journal
of the American Medical Women's Association 52(1):33–38

Wertz DC, Fanos JH and Reilly PR (1994) Genetic Testing for children and adolescents: who de-
cides? JAMA 272(11):875–881

Wertz DC and Fletcher JC (1998) Ethical and social issues in prenatal sex selection: a survey of geneticists in 37 nations. Soc Sci Med 46(2):255–273

Wertz DC, Fletcher JC (1989) Ethics and Human Genetics A Cross-cultural Perspective.
Springer-Verlag, Heidelberg

Consensus and Variation among Medical Geneticists and Patients on the Provision of the New Genetics in Germany – Data from the 1994–1996 Survey among Medical Geneticists and Patients

Irmgard Nippert, Gerhard Wolff

The ability to diagnose genetic diseases has increased rapidly over the past two decades. Due to the various international human genome programs and projects the knowledge of the structure and function of human genes is steadily expanding and genetics is on its way of becoming an integral part of medical care.

Genetic tests have the ability to detect genetic disorders or to detect increased susceptibility and predisposition to them. But most genetic tests differ from other diagnostic tests in medicine. In healthy people genetic tests have the ability to predict risks of future diseases, seldom however does the predictability approach certainty. Often no independent test is available to confirm the prediction of a genetic test, only the appearance of the disease itself confirms the prediction. For most genetic disorders no clinical interventions are yet available to prevent, treat or improve the outcome of future diseases. For other genetic diseases, the intervention trials that are underway have not yet proven save and effective treatment.

The results of genetic tests provide information not only relevant to the individual person but also relevant to the future health of relatives. The results of genetic tests may confront prospective parents with difficult options to consider in regard to reproduction.

People identified as at risk of disease or having future children with a genetic disease may experience psychological distress, social discrimination and stigmatization. Therefor transmitting genetic information from provider to consumer requires attention to the unique ways in which people may be affected by genetic information.

The following data report the outcome of a study that was undertaken in Germany to assess attitudes and values of consumers and providers towards ethical aspects of genetic service provision.

An anonymous questionnaire including "choice of action" questions in response to given situations covering among others topics such as:
- confidentiality versus duties to inform relatives at risk
- privacy of information from institutional third parties
- provision of predictive testing

- testing of children and minors
- selective abortion
- non-medical uses of genetic tests
- patient autonomy
- supporting client's decision

was distributed in 1994 – 1996 among all geneticists working in medical genetics in Germany and among patients undergoing genetic counseling at the Institut für Humangenetik, Westfälische Wilhelms-Universität, Medical School, Münster and at the Institut für Humangenetik und Anthropologie, Albert-Ludwigs-Universität, Medical School, Freiburg, Germany.

The study is part of the international study "Ethical Issues in Genetics", designed by Dorothy Wertz (The Shriver Center, Waltham, MA, U.S.A.) and John Fletcher (University of Virginia School of Medicine, Charlottesville, VA, U.S.A.) and was conducted in 37 nations worldwide. In Germany the study was funded by the Deutsche Forschungsgemeinschaft, internationally the study was supported by the Ethical, Legal and Social Issues (ELSI) program of the U. S. Human Genome Project (National Institutes of Health).

In Germany 259 geneticists and 591 patients responded. The response rate for geneticists was 65% and 67.5% for patients (see Table 1).

The mean age of the patients was 33.1 years, 78.5% were married and 47% had no children. The majority of the patients were female (90%). This high percentage of females reflects the fact that in genetics it is women that are mostly counseled for genetic risks and at whom genetic information and tests are mostly targeted.

The mean age of geneticists was 43.3 years, 70.2% were married and 30% had no children. Of the respondents 52.1% were female and 47.9% male. The study group of geneticists had spend an average of 11.1 years in genetics. Thus a very well expe-

Table 1. Sample characteristics

	Geneticists n = 259	patients n = 591
response rate	65%	67.5%
mean age	43.3 yrs	33.1 yrs
female	52.1%	90%
male	47.9%	10%
married	70.2%	78.5%
no children	30%	47%
~ years working in medical genetics	11.1	
~ % if time spend in medical genetics related to direct patient care and patient services	76%	

rienced group of medical geneticists is represented in the sample. The average percentage of time the geneticists spend in medical genetics that is related to direct care and patient services was 76%.

Consensus and Variation on Patient Autonomy in Genetic Service Provision

The duty to respect the self-determination and freedom of choices of competent persons as well as to protect persons with diminished autonomy (e.g. persons with severe mental retardation, young children) is one of the core principles for medical practice and research. An important way to ensure patient autonomy in regard to genetic testing is to provide adequate information upon which a person can make an informed decision whether or not he or she wants to undergo testing. Adequate genetic counseling enables an individual or family to make their own decisions after a process of gaining understanding of their personal needs, values and expectations. Optimum counseling enables an individual or a family to reach their own decisions about testing, early diagnosis, prevention and reproduction.

In Germany geneticists seem more paternalistically minded in their approach towards the provision of services than their patients wants them to be, especially in regard to access to services and to genetic information. Statistically significant disagreement was found in 11 out of 14 statements (see Table 2).

Whereas 47.6% of the patients thought that they are entitled to whatever services they ask for as long as they can pay out-of-pocket, only 12.8% of the geneticists would subscribe to that statement. 66.4% of the patients thought that withholding any services is a denial of the patients' rights, only 28.4% of the geneticists agreed. 93.5% of the patients thought that genetic tests should be available to all women who request them (90% of the respondents are female!) but only 23.1% geneticists agree.

Apparently German geneticists agree that patient autonomy in regard to genetic services should be limited in some ways. They believe that patients should have a right not to know (89.2%), though patients seem to be not so certain about this right, only 35.4% are agreeing. This difference may also be due to the possibility that geneticists are much more familiar with the concept of the "right not to know" than patients who may never have heard about it.

Nonetheless patients (86.1%) and geneticists (81.4%) believe that other people have a right to services of which their doctors may personally disapprove.

Table 2. Patient autonomy

	% agreeing		
	Providers	Patients	χ^2
1. Good genetic counselors should provide sympathy and understanding even if they disagree with a patient's decision	74.1	92.2	s.
2. After taking a test, people should have the right not to be told the results	89.2	35.4	s.
3. Tests on unborn babies should be available to all women who request them	23.1	93.5	s.
4. People are entitled to whatever services they ask for, if it is legal and they can pay for it out-of-pocket	12.8	47.6	s.
5. Withholding any services is a denial of the patient's rights	28.4	66.4	s.
6. Parents should be told the sex of the unborn baby if they ask	15.2	91.3	s.
7. Parents should not be told the sex of an unborn baby	34.1	3.9	s.
8. A doctor who refuse to do sex selection should offer to send patients to someone who would do it	6.4	28.7	s.
9. A woman's decisions about abortion should be her own, without intervention by anyone	63.2	51.8	s
10. The father of an unborn baby should be able to force the mother to have the baby tested to see if it is normal	5.6	19.6	s.
11. If the law forbids something in the one state or province, the doctor should offer to refer patients to a doctor somewhere where it is legal	17.0	29.0	s.
12. Parents should be told all test results about the health of the unborn baby	94.4	98.5	n. s.
13. When people ask for something that is legal but that their doctor is unwilling to do for moral reasons, the doctor should offer to send them to another doctor	81.4	86.1	n. s.
14. Parents should have the right to choose the sex of their children	0.8	3.9	n. S.

Level or significance ≤ 0.05

Sharing Genetic Information

Disclosure and confidentiality of test results represent one ethical dilemma occurring in clinical genetic practice. Confidentiality means a duty of the physician not to reveal information against a patient's wishes. In genetics the true patient may be a family with a shared genetic heritage (Berg 1989). Some geneticists argue that therefor family members have a moral obligation to share genetic information with each other (Berg 1994). It is argued that this obligation arises from kinship bonds and the principle of non-maleficience. Wertz, Fletcher and Berg state that "it is the counsellee's moral obligation to tell relatives at risk about a diagnosis and/or result of presymptomatic tests, so that these relatives can choose whether to be tested themselves. It is also a counsellee's moral obligation to provide blood, saliva samples or other specimens, so that relatives can have genetic tests" (Wertz, Fletcher, Berg, 1995).

In Germany no consensus was found among patients and geneticists on the issue whether or not a patient's relatives should be informed against the wishes of that patient (see Table 3).

Whereas the majority of the geneticists (58.6%) would maintain the patient's confidentiality and not tell the relatives, 29.4% would tell the relatives if they ask, and 11.2% would pass the problem to the referring doctor, only 23.8% of the patients want the geneticists to preserve confidentiality.

Patients also have a different view than geneticists of the rights of spouses or partners or blood relatives to have access to stored DNA without a person's consent (see Table 4). Geneticists are much more restrictive. Only 16.1% of the geneticists versus 31.9% of the patients would grant blood relatives access to stored DNA without consent and only 17.8% of the geneticists would grant access versus 42.3% of the patients who would do so.

Table 3. Disclosure to relatives, against patient's wishes

A patient diagnosed with Huntington diseases refuses to permit disclosure of relevant genetic information to relatives at risk.

| | % would (Providers) % providers should do (Patients) | | | |
	Respect confidentiality	Tell relatives, if they ask	Tell relatives unasked	Let referring doctor decide
Providers	58.6	29.4	0.0	11.2
Patients	23.8	41.5	34.7	not asked

Table 4. Access to stored DNA without person's consent

	% agreeing							
	Law enforce-ment agencies	Life insures	Blood rela-tives	Em-ployers	Health insurers	Spouse or partner	Employers for public safety	Registry of Motor vehicles
Providers	33.9	1.2	16.1	0.4	1.2	17.8	11.8	2.0
Patients	9.7	1.0	31.9	0.3	5.6	42.3	18.9	3.1

Table 5. DNA fingerprinting

	% agreeing		
Should be required for	Providers	Patients	χ^2
persons **convicted** of sex crimes, such as rape or child molesting	85.5	93.6	s.
persons **convicted** of other crimes	81.0	83.2	n. s.
persons **charged** with sex crimes	89.1	82.3	n. s.
persons charged with other serious crimes	83.5	70.6	s.
members of the armed forces, so that war dead could be identified	35.4	61.9	s.
newborns, to prevent mixing up babies in the hospital	11.7	31.2	s.
applicants for passports, to prevent fraud	4.0	10.4	s.

Level or significance ≤ 0.05

Surprisingly German geneticists are much more in favor of granting access for law enforcement agencies (33.9%) than patients (9.7%).

On the other hand both most geneticists and patients responses about DNA fingerprinting (see Table 5) were positive. Large majorities agree that DNA fingerprints should be required and kept.

Prevention of "Genetic Defects"

In genetics today prevention is supposed to differ from any eugenic approach that favored the improvement of the "population quality" and that supports third parties' intervention with people's reproductive rights, because it is reasoned today genetics is primarily based on supporting individual and family choices. But choices, especially reproductive choices, are not made in a social vacuum (Wertz, 1996). A lot of socio-economic and socio-cultural norms and factors, such as women's roles in society, income, family size, cultural expectations, availability of resources for the disabled, availability of reproductive techniques all exert influence on individual choices. A substantial majority of the patients (79.5%) thought that a woman should have tests on the unborn baby if she is at risk of having a child with a serious disease or disability and a third (33.5%) of the geneticists agreed as well. If the amount of agreement found in the two study populations is compared, it clearly shows that patients are much more inclined than geneticists to support the idea that people have a duty to prevent birth defects or that people have a responsibility not to pass on serious diseases or disabilities to their children (see Table 6).

Geneticists and patients seem to take a fairly gloomy view of disability. Most agree that our society will never provide enough support for people with disabilities (see Table 7) and that some disabilities will never be overcome, even with social support.

Table 6. Prevention

	% agreeing		
	Providers	Patients	χ^2
It is not fair to a child to bring it into the world with a serious genetic disease or disability if the birth could have been prevented	17.8	47.6	s.
Women are under a lot of social pressure to have tests on their unborn babies	48.4	28.9	s.
A woman should have tests on the unborn baby if she is at risk of having a child with a serious disease or disability	33.5	79.5	s.
Before marriage, responsible people should find out whether they could pass on serious diseases or disabilities to their children	23.4	48.7	s.
People at high risk for passing on serious diseases or disabilities to their children should not have children	13.2	40.3	s.
Two people who could pass on a serious condition to their children if they marry each other should not marry each other. They should try to find another partner with whom they could have healthy children	3.2	3.9	n. S.

Level or significance ≤ 0.05

Table 7. Perceptions of Disability

	% agreeing		
	Providers	Patients	χ^2
Our society will never provide enough support for people with disabilities	56.9	58.8	n. s.
Some disabilities will never be overcome even with social support	91.8	57.7	s.
Persons with severe disabilities make society more rich and varied	38.2	20.9	s.
It is not fair to a family's other children to give birth to a child with a disability, if the birth could be prevented	10.0	20.8	s.
Parents should always have counseling before tests about an unborn baby	96.0	98.0	n. s.
It is useless to perform tests on an unborn baby when there is no treatment for the unborn baby	6.8	18.3	s.
It is socially irresponsible knowingly to bring an infant with a serious disability into the world	7.6	26.7	s.

Level or significance ≤ 0.05

Since more than 25 years prenatal diagnosis (PD) has opened up the possibility to selectively abort fetuses with genetic conditions. The number of conditions that can be detected prenataly is constantly increasing and is now possible for more than 600 conditions (Weaver, 1992), some of which involve only mild clinical symptoms. A majority of geneticists reported that they personally would abort for 7 out of 14 conditions (see Table 8.), a majority of patients would abort for 8 out of 14 conditions. In Germany 71.0% of the geneticists would abort for Trisomy 21 (68.5% of the patients), 89.9% for severe, open spina bifida (80.6% of the patients), and 60% for Cystic Fibrosis (63.5% of the patients).

Agreement in the personal attitudes towards abortion among patients and geneticists was found in 8 out of 14 conditions. Towards 4 conditions: Huntington, severe obesity, 45,X and Achondroplastic dwarfism patients, 90% of whom were women, took a more conservative view toward abortion than did geneticists.

The majority of geneticists and patients do not tolerate the idea of sex selection and selective abortion after PD. They both (90.3% geneticists/87.6% patients) agree that genetic services should *not* provide this option and that sex selection should be made illegal. These data clearly support the view that sex selection is not culturally acceptable in Germany. Because of personal and cultural objections to use PD for sex selection and to abort for this purpose it is highly likely that re-

Table 8. How would you personally respond if you faced the possibility in the first trimester of gravity of having a child with the disorder listed below

	I would have an abortion		I would not have an abortion but it should be available for others		I would not have an abortion and it should be illegal for others		
	% agreeing		% agreeing		% agreeing		
	Providers	Patients	Providers	Patients	Providers	Patients	χ^2
Anencephaly	95.8	93.9	2.5	5.1	1.7	1.0	n. s.
Trisomy 13	94.5	88.3	5.1	9.0	0.4	2.6	n. s.
Severe, open Spina bifida	89.9	80.6	9.2	15.7	0.8	3.7	s.
Huntington Disease	47.9	33.3	44.5	42.9	7.6	23.7	s.
Trisomy 21	71.0	68.5	26.5	22.8	2.5	8.7	n. s.
Cystic Fibrosis	60.9	63.5	36.0	27.1	2.6	9.4	s.
Severe obesity in absence of a known genetic syndrome	19.2	9.3	51.1	39.8	29.7	50.9	s.
PKU	24.9	72.2	60.1	22.4	15.0	5.4	s.
45,X	18.5	8.9	63.1	47.2	18.5	43.9	s.
XXY	18.1	32.8	64.1	43.3	17.7	23.8	s.
Hurler Syndrome	87.4	87.6	11.7	9.8	0.9	2.6	n. s.
Achondroplastic dwarfism	54.1	28.3	40.3	51.1	5.6	20.6	s.
Cleft lip and palate in a girl	6.8	4.7	39.1	43.4	54.0	51.9	n. s.
Cleft lip and palate in a boy	7.1	5.1	40.8	38.4	52.1	56.5	n. s.
Child is not the sex desired by the parents	0.4	0.3	9.3	12.1	90.3	87.6	n. s.
Sickel cell anemia	39.1	65.8	52.2	27.6	8.7	6.6	s.
Predisposition to alcoholism	3.4	4.6	52.3	43.8	51.7	44.3	n. S.

Level or significance ≤ 0.05

Table 9. Non-medical uses of prenatal diagnosis: Prenatal paternity testing (nonforensic).

A pregnant woman requests tests to find out who the baby's father is. She is involved with two men: Joe, who wants children, and Bill, who does not. If Joe is the father, she will have the child. If Bill is the father, she will not have the child. If she cannot find out which man is the father, she will probably not have the child. Only one man needs to be tested and Joe has offered to cooperate

% would (providers)	% providers should do (patients)	
	Providers	Patients
Do the test without comment on her personal situation	1.2	9.0
Do the test with some comment on possible consequences of her actions	7.1	26.9
Do the test only if she agrees to counseling	3.2	46.5
Refuse to do the test	74.7	17.6

quests for PD for sex selection will remain few. The major use for sex selection will occur in nations and cultures where there is a strong preference for sons, as for instance in India and China.

Aside from sex selection in the absence of an X-linked disorder, nonforensic prenatal paternity testing represents another non-medical use of PD. Women may request PD solely for paternity testing in cases where paternity is uncertain. The German Society of Human Genetics has clearly stated opposition towards this use.

However the majority of patients, take a different view. Whereas 74.7% of the geneticists would refuse to perform the tests more than 80% of the patients would expect the test to be done either with or without comments and prior counseling.

Testing Children for Late-onset Disorders

Some genetic tests indicate whether a person will develop a genetic disorder in the future. Tests for Huntington's disease, a mono-genetic disorder, offer a nearly hundred percent certainty but do not tell at what age the disease will appear. Other tests offer percent life-time risks that a disease may occur. The decision whether or not to undergo predictive or presymptomatic testing for late-onset disorders (=disorders that become apparent only in middle age or later) may be influenced by a number of factors such as: personal knowledge of the disease, interpretation of risks, whether or not a condition is treatable or preventable. It is widely agreed upon in the geneticists' scientific community that proper informed consent has to be obtained from any person who is considering genetic testing: "The person should be given information about the risks, benefits, efficacy, and alternatives to

Table 10. Testing children for late-onset disorders

% who thought parents should be able to have children < 18 tested for

	% agreeing		
	Huntington disease (HD)	Cancer genes	Familial hypercholesterolemia
Providers	13.7	45.1	55.1
Patients	36.9	85.8	87.2

the testing; information about the severity, potential variability, and treatability of the disorder being tested for; and information about the subsequent decisions that will be likely if the test is positive" (Andrews et. al, 1994).

Testing children for late-onset disorders in the absence of medical benefits on parental requests has been widely debated. The German Society of Human Genetics opposes predictive testing of children and minors unless there is a clear benefit such as prevention for the minor (Kommission für Öffentlichkeitsarbeit und ethische Fragen der Gesellschaft für Humangenetik e. V., 1995).

This attitude clearly reflects the increasing respect for minor's autonomy in the overall context of voluntariness of genetic information and genetic testing minors should have the right or not knowing their genetic state if they wish so.

Most geneticists in Germany oppose testing children for a untreatable disorder such as Huntington's disease, however more geneticists are inclined to approve predictive testing if there is a chance that early intervention may improve the outcome of the disease (see Table 10).

Substantially more patients than geneticists think that parents ought to be able to have their children under 18 years of age tested for late-onset disorders.

In Conclusion

Geneticists in Germany are less willing to support patient autonomy than patients wish them to do. The restrictiveness the geneticists showed by their answers may be due to a deep felt fear of potential misuse (as for instance for sex selection) of genetic tests and the fear to be blamed for this or to be partly held responsible for misuse.

Geneticists are less willing to test children for late onset disorders than patients. Here geneticists are in accordance with the emerging international guidelines on this issue (WHO, 1998; Andrews et. al, 1994). This issue has been more debated among specialists than in the public therefor it is of no surprise that patients are still more in favor of testing children in the absence of a clear medical benefit.

Patients are more ready to share genetic information and to allow access to information for spouses and blood relatives without a person's consent, whereas geneticists stick to a much more strict protocol and are acting in accordance with a patient's wishes.

Patients also are more likely to support non-medical uses of genetic tests such as nonforensic paternity testing than geneticists. But both, patients and geneticists, agree in their opposition towards sex selection.

It is interesting that agreement was found in regard to personal attitudes towards selective abortion in the majority of the represented cases. This agreement can be interpreted to be an indicator for culturally shaped perceptions of disability and shared personal values and norms.

The disagreement to be found among geneticists and patients on how genetic diseases should be prevented, shows a more coercive attitude in patients towards persons at risk of having a child with a genetic condition. Patients are much more inclined to support the idea of "responsible parenthood" than geneticists who would rather support the individuals' right to exert the freedom of their reproductive choices.

The data demonstrate that there is a need for more public discourses on the ethical and social implications of applied genetics.

References

Andrews, L. B.; J. E. Fullarton, N. A. Holtzman, A. G. Motulsky (1994) Assessing Genetic Risks, Implications for Health and Social Policy, National Academic Press, Washington

Berg, K (1989) Preface. In: Wertz DC and Fletcher JC (eds.) Ethics in Human Genetics: A Cross-Cultural Perspective, Springer Verlag, Berlin

Berg, K (1994) The need for laws, rules and good practices to secure optimal disease control. In: Ethics and Human Genetics. Proceedings of the 2nd Symposium of the Council of Europe on Bioethics, 30 November – 2 December 1993, Council of Europe Press, Strasbourg, pp. 122-134

Kommission für Öffentlichkeitsarbeit und ethische Fragen der Gesellschaft für Humangenetik e. V. (1995): Stellungnahme zur genetischen Diagnostik bei Kindern und Jugendlichen. Med Genetik 7:358-359

Weaver, D. D. (1992) Catalog of Prenatally Diagnosed Conditions, The Johns Hopkins University Press, Baltimore

Wertz, D. C., J. C. Fletcher, K. Berg, (1995) Guidelines on Ethical Issues in Medical Genetics and the Provision of Genetic Services, World Health Organization, Hereditary Diseases Programme, Geneva

Wertz, D. C. (1997) Society and the Not-so-new Genetics: What are we afraid of? Some Future Predictions from a Social Scientist. The Journal of Contemporary Health Law and Policy, Vol. 13, 299-346

World Health Organization (1998) Proposed International Guidelines on Ethical Issues in Medical Genetics and Genetic Services, Report of a WHO Meeting on Ethical Issues in Medical Genetics, 15 – 16 December 1997, Geneva

The Case for Proposed International Guidelines on Ethical Issues in Medical Genetics

John C. Fletcher, Dorothy C. Wertz

Introduction and History of Guidelines

The proposed international guidelines for issues in medical genetics have a history that we sketch here before moving to the main question to be addressed in this paper: what is the case for these guidelines? In addressing this question, we also respond to several others, posed by the organizers of the conference. Why were the guidelines developed? What ethical principles underlie the guidelines? Even if widely adopted, what kind of moral authority can international guidelines have in the face of challenges in the name of cultural and moral diversity?

The proposed guidelines grow out of a dialogue in which we have participated since 1994 within the Hereditary Diseases Program of the World Health Organization (WHO). The results of that dialogue were published in a monograph in 1995 [1] that concluded with the proposed guidelines. It is important to note that the monograph is not an official document of the WHO and represents only the views of its four authors. However, the document was extensively reviewed by members of the WHO Expert Advisory Panel on Human Genetics, seven additional reviewers, and members of the WHO Secretariat.

Prior to our current appeal to medical geneticists to act through the appropriate international forum to debate and adopt a set of guidelines for their collective professional conduct, we have conducted two international studies of the ethics of genetics professionals. The first, in 19 nations, was done in 1985-86 [2]; the second, covering 37 nations (including the 19 from the first survey) is complete. Results and discussion of the second survey will appear in professional journals and a book similar to one published following the first survey [3]. Both surveys were conducted through anonymous questionnaires describing most of the ethical problems that occur in the practice of medical genetics, usually presented in the form of case vignettes. We included all nations with ten or more practicing medical geneticists.

Why were the Guidelines Developed?

The rationale both for the study of ethical problems in medical genetics and for the proposed guidelines is that science and the ethics of professionals do not stop at international borders. We know empirically that geneticists around the world face similar problems as healthcare professionals. As members of a profession, there is an obligation for self-regulation in the arena of professional ethics. Depending upon the particular issue, there is greater or lesser consensus among the world's geneticists. Our first international study found more divergence of ethical views among geneticists than convergence. The second study enlarged upon certain themes and found a large degree of diversity on some issues but moderate to strong consensus on others.

There are other data indicating that geneticists may be open to guidelines for professional conduct. Most geneticists agree that if a genetic service was unavailable or illegal in their own country, they should offer patients a referral across international borders. In this event it would make sense to have some international agreement about what should be available and how it should be provided. Discussion can only proceed in the light of full awareness of international perspectives.

What Ethical Principles Underlie the Guidelines?

There are eleven issues addressed by the guidelines. We have a high degree of confidence, based upon our international studies, that these are the ethical issues that most frequently face medical geneticists around the world and that their responses to these issues could be embodied in a set of guidelines that can be further grounded in ethical principles having wide appeal and moral scope across and beyond cultural and international boundaries.

The issues are:
1. access to services,
2. directive versus non-directive genetic counseling,
3. voluntary versus mandatory provision of services,
4. full disclosure to counsellees,
5. confidentiality versus the obligation to inform relatives at genetic risk,
6. protection of privacy of genetic information from institutional third parties,
7. prenatal diagnosis
8. protection of choices relevant to genetic services,
9. rights of adoptive children and children conceived from donor gametes,
10. research issues, and
11. experimental human gene therapy.

Individuals and groups must choose their sources of ethical guidance. The basic ethical principles that underlie the guidelines are embedded in traditions and institutions in every society. In pluralistic societies, no one ethical tradition dominates, but each tradition has a place for basic principles. These principles have also been sources for appeal and moral deliberation in several other national, international, and scholarly considerations of ethics in research, in medical practice, and in human genetics services. [4]

Respect for persons: the duty to respect the self-determination and choices of autonomous persons, as well as to protect persons with diminished autonomy (e.g. young children, persons with mental retardation, and those with other mental impairment). Respect for persons includes fundamental respect for the other; it should be the basis of any interaction between genetics professional and counsellee.

Beneficence: the obligation to secure the well-being of persons by acting positively on their behalf and, moreover, to maximize the benefits that can be attained.

Normaleficence: the obligation to minimize harm to persons, and, wherever possible, to remove the causes of harm altogether.

Proportionality: the duty, when taking actions involving risks of harm, to so balance risks and benefits that actions have the greatest chance to result in the least harm and the most benefit to persons directly involved and to others who will be affected by the specific choices involved.

Justice: the obligation to distribute benefits and burdens fairly, to treat equals equally, and to give reasons for differential treatment based on widely accepted criteria for just ways to distribute benefits and burdens.

Cross-Cultural Issues, Ethical Relativism and the Ethical Perspective in the Guidelines

There was much discussion for and against the concept of an international code of ethics at the 8th International Congress of Human Genetics in 1991. We published an account of this discussion and our responses to the objections to a code in 1993. [5]

In our view, among the most serious objections to a proposed code were:

1. a code would exclude the views of those whose opinions reflect cultural minorities and would therefore be oppressive, and
2. ethics is too subjective and relative to cultural determinants to codify. The issue of cultural differences in moral perspectives and of ethical relativism cannot be

ignored, especially in the context of proposing international guidelines for professional ethics.

The perspective that we have adapted for our cross-cultural research and that we believe to be expressed in the proposed guidelines involves respect for moral opinion and practices found in different cultures and also in various sectors of a pluralistic society. But this respect is not unlimited. There are practices that the guidelines would not condone.

Accordingly, the guidelines represent flexibility of moral position in some cases, e.g., „genetic counseling should be as non-directive *as possible* (thus recognizing some cultural barriers and habituation to paternalistic practices that could be gradually remedied by education and training). On other issues, such as using prenatal diagnosis for other than reasons relevant to the health of the fetus, e.g., for gender selection, the guidelines would be violated by such a practice.

A second controversial moral position embedded in the guidelines is to encourage the societal protection of choices, including the choice of a genetic abortion following a prenatal diagnosis with positive findings; an assumption behind the guidelines is that data ought to be collected regarding the benefits and burdens of social policy and laws that protect or proscribe such choices. The guidelines should rightly be interpreted as encouraging the protection of the freedom to choose or refuse genetic services without discrimination or stigmatization. In nations where such choices are legal, and the right of referral outside of a nation where such choices are proscribed.

International guidelines for genetic processionals must be framed by an ethical perspective that responds to cultural differences, on the one hand, and on the other hand draws lines at moral limits and rules out the intolerable-ones must be extremely wary of sanctioning coercion in the name of morality. But it can also be the right thing to do and lead to beneficial consequences. For example, it was surely not morally wrong in 1829 for the Britisho Governor General of India, Lord William Bentinck, [6] to abolish and use coercion to stop the practice of suttee, the custom of burning widows alive with the corpses of their husbands. The justification of the use of force to enforce morality depends upon the circumstances and the proportionality of the harm to be overcome by the coercion to the harm done to received moral practices by coercion itself. We do not make this judgment as a statement of objective moral truth. Judgments about coerciveness in the name of morality must be made in the context of specific but changing circumstances and with a careful eye towards the consequences such judgments are relative to the circumstances, which include cultural differences. In India, government officials and physicians have had to turn to the law again to try to abolish practices of wasteful use of the scarce resource of prenatal diagnosis for sex choice alone. In our view, this law is morally justified, because of the greater good that could be done with these resources directed towards genetic diseases in that nation, as well as to avoiding abortions of normal female fetuses for gender reasons alone.

Conclusion

To recapitulate and conclude, seeking guidance about professional ethics in human genetics cross-culturally involves three steps that we have taken with many colleagues around the world:

1. to describe accurately the major ethical problems in the activity under study,
2. to conduct accurate studies of consensus and variation in prevailing approaches to ethical problems and of the consequences of these approaches and
3. to evaluate the prevailing set of approaches using a perspective in ethics such as the composite theory described above.

The fourth step in guidance-seeking is to recommend reforms in prevailing professional approaches to the ethical problems of human genetics on a local, national, and global level. We now recommend the use of the term „guidelines" rather than „code" to embody a more flexible approach but one that is committed to advances and practices that have proven benefits for patients and the genetic professionals who serve them.

References

1. Wertz DC, Fletcher JC, Berg K, Boulyjenkov V (1995) Guidelines on Ethical Issues in Medical Genetics and the Provision of Genetics Services. World Health Organization. Hereditary Diseases Program, 1995. [WHO document, WHO/HDP/GL/ETH/95.1, 1995. Single copies are available, on request, from Human Genetics Program, WHO, 1211 Geneva 27, Switzerland] An additional shorter version has been published „Proposed International Guidelines on Ethical Issues in Medical Genetics and Genetics Services." Report of a WHO Meeting on Ethical Issues in Medical Genetics. World Health Organization, Human Genetics Program, Geneva, 1998 [Single copies are available, on request, from Human Genetics Program, WHO, 1211 Geneva 27, Switzerland]
2. Wertz DC, Fletcher JC, Mulvihill JJ (1990) Medical geneticists confront ethical dilemmas: cross-cultural comparisons among eighteen nations. American Journal of Human Genetics 46(6):1200-1213
3. Wertz DC, Fletcher JC (1989) Ethics and Human genetics: A Cross Cultural Perspective. Springer-Verlag, Heidelberg
4. United States 1983 President's Commission for the Study of Ethical Problems in Medicine and Biomedical and Behavioral Research. Screening and counseling for genetic conditions. US Government Printing Office, Washington DC; Fletcher JC, Berg K, Tranoy KE (1985) Ethical aspects of medical genetics: a proposal for guidelines in genetic counseling, prenatal diagnosis, and screening. Clin Genet 27:199-205; CIOMS (1993) International ethical guidelines for biomedical research involving human subjects. Council for International Organizations of

Medical Sciences, Geneva; Beauchamp TL, Childress JF (1994) Principles of biomedical ethics. Oxford University Press, New York, 4th ed
5. Wertz DC, Fletcher JC (1993) Proposed: an international code of ethics for medical genetics. Clin Genet 44:37-43
6. Hare RM (1981) Moral thinking. Clarendon Press, Oxford 25-43
7. Verma IC, Singh B (1989) Ethics and medical genetics in India. In: Wertz DC Fletcher JC (eds) Ethics and Human Genetics: A Crosscultural Perspective. Springer-Verlag, Heidelberg p 267

Session IV

Critiques on the Provision of the New Genetics: Pros und Cons from Consumer and Provider Perspectives

Statement on CF-Heterozygote Testing

STEFAN KRUIP

As adults living with CF, we feel deeply concerned about the new possibilities in the field of genetic diagnoses in cases of CF, and are considering how to deal with these new techniques. Up to now genetic advice and diagnosis in the case of CF were reserved for people directly or indirectly affected. For example: The brother of a CF patient wanted to know the risk for his children, or a cousin had her pregnancy prenatally tested. But instead of helping CF families clear up the problematic cases, the heterozygote testing enables experts to search for people with CF by testing a great number of people who have an average probability of being a CF gene carrier.

So far these tests are difficult and expensive to make. That is why we should be sure what the aim of the heterozygote testing of a great number of people is to be.

Only couples who are both gene carriers can expect to have a CF child. So these couples are looked at first. Then there is the possibility of testing all the newborn children of these couples in order to find the CF patients quickly, and to give them the best medical care from the start. But this aim, it must be said, could be achieved with much less effort and expense, if the reliable trypsin test were used for every newborn child, and if pediatricians, internists, and specialists for pulmonary diseases were better acquainted with the symptoms of CF and therefore enabled to diagnose CF more easily.

Very often it is mentioned that the aim of these tests is to give parents more options to act: that means that on the basis of this additional knowledge, parents may come to decisions which formerly were not possible. Essentially this means a decision of life and death of the unborn child with CF in case of a positive result of the prenatal diagnosis. In the individual case, when a mother's burden seems intolerable, (for example, if she already has a child with CF) the parents' personal decision must naturally be respected. But if these tests are offered to the population as a whole, tests which may result in finally avoiding all CF patients, then we must realize the intention of this proceeding.

Those who recommend abortion mostly mention of the following reasons:

1) Parents cannot be expected to carry the burden of a sick child.

Besides the fact that abortion may be a heavy burden for the mother, this statement makes pregnancy dubious in general. 96% of the handicapped have been handicapped only because of accidents and disease in the course of their lives. Parents are probably more burdened by mental, social, or financial problems than by CF.

2) The child's suffering can be avoided by abortion.

As adults who live with CF, we do not approve of this strategy: suffering avoided - patient dead. We think that it is completely incomprehensible and inappropriate to leave this decision to a doctor. Those who want to terminate pregnancy in cases of CF must be asked: 1) How can you decide whether an unborn child with CF wants to live or should rather not be born? 2) How do you measure the value and sense of life? 3) Don't you think that difficulties and knowledge of one's own limits offer the chance of a more intensive life in comparison to the lives of careless so-called nonhandicapped? To quote the former Bundespräsident Richard von Weizsäcker: „If social behavior were the standard example, we should emulate people with Down syndrome. Compared to the sensibility that enables people who are deaf and blind to perceive through their skin, it is those able to see and hear who are disabled. Perhaps for a wheelchair user, a professor who can neither laugh nor weep seems to be disabled in his humanity." (Main-Post, 10.07.1993).

3) The strategy of „screening and abortion" is better from the economical point of view because it is cheaper than treating the sick.

It has already been proved that this assertion is wrong and that the expenses of avoiding a new case of CF amounts to more than 1 million dollars. I am glad to say that in Germany, we do not know anyone who upholds this unethical reason for heterozygote testing.

The possibility of the CF heterozygote testing constitutes quite a number of dangers which we want to explain in short:

CF may be avoided. Persons who are well informed can take advantage of the test and thereby ensure to get [sic] a „healthy" child or, at worst, terminate the pregnancy. In society the expectation of getting healthy children increases, and it will be more difficult to refuse these tests and to stand up for one's possibly handicapped children. Do the parents really have an additional option of acting if even the gynecologist gives the advice: „Ask for human genetic counseling so that you can expect a healthy child." If there is a positive diagnosis, abortion will automatically be result.

Parents must decide on life or death for their children under pressure of time, and without sufficient knowledge. Which average person really knows how people with CF live? Who knows that most of those afflicted by CF appreciate living and disapprove of abortion as a „solution," and that they see their disease as a challenge to live a desirable life?

As soon as wide sections of the population are interested in the CF test, the capacity of the human genetic advice centers will be insufficient. In this case, the testing will disassociate itself from the human genetic counseling, and it will be offered commercially. Parents will be left absolutely alone with the diagnosis, and with the pressure to come to a decision that can hardly be justified either way.

Even without the screening program and its introduction as a standard test, the CF test acquires a eugenic nature as soon as the sequence gene carrier test - prenatal diagnosis - abortion has been established and society accepts this in order to avoid the burden for the parents and the health care system.

During the course of the conference „From Dialogue to Joint Action" in Bad Herrenalb in September 1993, we were relieved to notice that many human geneticists share our uneasiness and doubts; see for example „Statement on Prenatal Diagnosis and Abortion" by the Committee for Public Relation and Ethical Questions of the Society for Human Genetics e. V. (Ethik der Medizin 1992, 4:199-201) and „Statement on a Feasible Heterozygote Testing in the Case of CF" by the professional association of Medical Genetics (Medizinische Genetik 1990, 2/3:6).

We agree with our doctors that the diagnosis of CF cannot generally be succeeded by recommending abortion. Each year the statistical life prognosis increases by nearly one year. From the statistic point of view, a patient born in 1965 still had, at the age of 19 (1984), a life prognosis of three years. In 1991, he was 26 years old, and, according to the statistics, could expect to live four more years. At the moment, adults with CF experience that the older they become, the more their life expectancy increases - statistically. The dynamics of advances in different fields of treatment including first experiments with gene therapy in the case of CF imply that nobody is allowed to say anything definite about the life prognosis of a child born today. Adults with CF work, marry, and have children. Counseling must explain these positive developments to the parents, as well as the risk of a severe course and a short life.

In other European countries, CF heterozygote testing is not offered at all, or studies in this field have been stopped accordingly in Berlin and Göttingen. Pilot studies are carried through against the declared will of the CF self-support organizations. There, persons having a normal risk are offered the test as well as a leaflet containing information. Because of the reasons mentioned above, we disapprove of these studies and ask the experts involved to assume their responsibility for their actions. Advertising this test may be the beginning of the „brave new world" where being born becomes an embryonic steeplechase and where human genetics degenerates into a program to ensure quality.

Together with responsible human geneticists, we want the following requests to be generally accepted:

1. It should be required by law that genetic diagnosis be obtained only in combination with human genetic counseling before and after the testing.
2. The CF test should only be offered to persons seeking advice because they have an above average probability of being a gene carrier. This should happen before pregnancy occurs so that decisions are not made under pressure of time.
3. It must be extensively explained what life with CF is like, including the experiences of persons afflicted by CF and the advances in treatment. We support the employment of psycho-social counselors who should be obliged to work together with self-support groups.
4. A consensus that a general CF screening should not be introduced, and that pilot studies are of no value, and should be stopped.
5. Gynecologists should be asked to respect and support the principles of non-directive counseling, and to increase the parents' competence of a decision and of the independent decision of the parents.

The Patients' Perspective on the Provision of Genetic Testing

ALASTAIR KENT

Introduction

Scientific advance is uncovering the structure and function of genes which cause or which predispose us to serious disease on a daily basis. Every working day another one or two genes are mapped and their structure and function worked out somewhere in the world.

Once a gene has been characterized, it is possible to develop a diagnostic test which will determine the presence of specific mutations and in so doing predict with greater or lesser certainty the implications for the future health and well-being of the individual on whom the test has been carried out.

But testing for the presence of a genetic mutation is not the same as being able to cure the disease that the mutation causes or predisposes us to. The gap between the isolation of the gene and the development of a cure is a large one. Although substantial progress has been made in bridging this gap for a small number of diseases, for the vast majority of those affected or at risk, genetic disease remains incurable and this is likely to be the case for the foreseeable future.

So why do patients generally regard scientific progress and the development of diagnostic tests for genetic disease as 'a good thing', notwithstanding the fact that the emergence of genetic testing is not an unalloyed benefit and that it potentially carries with it the risk of discrimination through the inappropriate application of the results of genetic testing, particularly but not exclusively in non-medical contexts?

Screening and Testing

Before considering the advantages and disadvantages of the application of our emerging understanding of the contribution of genetics to the early detection of disease it is necessary to be clear of the difference between *screening* and *testing*.

Screening is the testing of whole populations or of specified sub-sections of a population known to be at increased risk (e.g. pregnant women, women over 50, people from particular ethnic groups) from the presence of a specific gene or genes. Apart from an increased risk resulting from membership of the sub-set of the population being screened, there is no information that might lead the screener or the screened to suspect that any given individual is more or less at risk than any other.

Testing is used when other information, known to the individual and his or her doctor leads to the determination of a raised risk in that individual or family for which DNA or other investigations will provide confirmation of the diagnosis.

In the remarks that follow I shall confine my attention to testing - i.e. the use of DNA diagnostic techniques in individuals and families where there is evidence of pre-existing risk.

Testing in Adults

Where science offers the possibility of a DNA based diagnosis there are a number of conditions and criteria that must be fulfilled if the acquisition of test results is to be enabling for the recipient and not disabling.

Genetic data, once known, cannot be un-discovered. Once a test has revealed a condition to be present pre-symptomatically or that one is a carrier or has a raised risk, this information forms a permanent aspect of the individual's persona. It also is not confined in its consequences to the individual, but has implications for his or her family - either directly, in that they may themselves be at an increased risk or, indirectly, in that they may face the prospect of becoming a carrier at some point in the future.

For this reason it is important that individuals *must* remain the final judge of whether or not they wish to be tested. No-one should be forced into a situation where they are obliged to discover something about themselves by a third party when they would not choose to have that information themselves. Neither the state nor commercial interests (such as insurance companies) have the right to re-quire someone to be tested except in very narrow circumstances where the presence (or absence) of a particular genetic configuration can be demonstrated to have a causal link to a significant third party risk. Such instances are extremely rare - the Nuffield Council on Bioethics, in its report into genetic screening, could only come up with sickle cell disease in airline pilots as an example.

However, the right of one family member not to know does not extend to include the power to prevent another member of the same family from establishing his or her own genetic status, even if the acquisition of that knowledge has implications for the first party.

Given all this, genetic testing ought to be carried out in a context where proper pre- and post-test counseling is available. Where people can explore the consequences of actions before they are taken and appreciate the implications for themselves and for their wider family when they know the outcome. Such counseling ought, in so far as it is possible, to be non-directive and families ought to be able to be confident that they will be able to make decisions and reach conclusions that will be valued and respected whether or not they fit in with the pre-conceptions or assumptions by professionals (or by society as a whole) as to what the 'correct' or 'responsible' course of action should be.

Why do people want to be tested, especially when there is no cure or effective intervention that can be made to delay or ameliorate the symptoms of a given genetic condition?

Clearly individual reasons for wanting this knowledge will be as many and varied as the individuals concerned, but underlying many of the specific issues quoted is the wish to understand and to take control over that which has, hitherto, been uncontrollable - seeming like the operation of blind chance.

Decisions about genetic testing are usually made by individuals after personal experience of the condition in the family. Individuals do not walk in 'off the street' and ask to be tested for a rare genetic disorder unless they know something about the disorder in question and have a reason for asking to be tested.

Reasons for testing may include the desire to plan for the future - to make arrangements for long term care in the case of Huntington's disease or familial Alzheimer's for example, to avoid pregnancy or to avoid continuing an affected pregnancy if one knows that one is at risk, or to permit interventions to be made in advance of symptoms emerging - as is possible with hypercholesterolaemia for example, where dietary adjustment, medication and exercise can maintain health and avoid disability.

However, testing to acquire knowledge in one area of life can create a situation that produces problems in another. For those identified as being presymptomatically at risk, obtaining insurance cover can be difficult. Particularly if the insurance company is not required to demonstrate relevance and competence in respect of the data require. It is both judge and jury in its own court as to the importance it gives to genetic information as a factor in the decision whether or not to grant cover.

To avoid abuse genetic testing must be regulated, both in respect of the technical and professional competence of those providing the service, the circumstances under which this service is made available to the public in general and to individuals and families at risk in particular. If code of practice and voluntary self regulation is seen to be ineffective, then legislation must be introduced to prevent abuse and eliminate unfair discrimination.

Testing Children

Where there is a clear therapeutic benefit that results from a genetic test (such as is the case with PKU testing in new-borns) then there is no argument for not carrying out (with parental consent) a diagnostic test.

In cases where there is no such therapeutic advantage the issue is much less clear cut. A recent report by the U.K. Clinical Genetics Society recommended that, in such cases, testing should not be undertaken until the child was of an age to play an informed part in the decision to be tested, as to do otherwise would compromise his or her autonomy.

The Genetic Interest Group surveyed its membership (over 100 support groups for those affected by specific conditions) and came up with slightly different conclusions. Against a background assumption about the availability of counseling, information and support, GIG's membership felt that the decision as to whether or not to test for carrier status in recessive conditions should, ultimately, be that of the parents - although it was stressed that it was desirable that the child should normally be fully involved in the decision and, for this reason, it is better to leave taking action unless there is good reason not to.

Ante-natal testing for late onset conditions should not be undertaken unless the mother has expressed a wish to terminate an affected pregnancy. Should she change her mind subsequently then the original intention cannot be enforced and no pressure should be brought to bear to try and do so. Antenatal testing for recessive conditions will reveal carrier status as well as affected fetuses but this is a small price to pay when weighed against the benefits - the 'reproductive confidence' that accrues from ante-natal testing.

Consent

Testing children raised the issue of consent. No testing should be done without informed consent of the individual, in the case of adults or the parent or guardian for those too young to give valid consent on their own account.

Those unable to give consent by reason of mental disorder, learning disability or for some other reason are a special case. Normally the issue of personal benefit is sufficient to decide the issue - as has been established in the Council of Europe's bio-ethics convention. But in rare cases this may not be an adequate criterion. Where DNA tests do not exist yet, or where complex genes with many mutation possibilities control the condition, studies involving extended families may be the only way of establishing risk. Again a survey of GIG's membership indicated that, against a background where most people give their consent when asked to do so, there should be a presumption of willingness on the part of those unable to consent, rather than an assumption of unwillingness. However this can only be applied under defined circumstances where there is no other way of getting the information and it is only justifiable given the minimally invasive process of obtaining a blood sample. The logic could not be extended to support kidney donation, say, or some other procedure with significant risk to the individual.

It has been suggested that a focus on ante-natal testing, with the consequent offer of termination of pregnancies diagnosed as affected, will result in a diminution of resources available for those born with disabilities, and the possibility of stigmatization of those who are born with a genetic disorder.

In the developed world increasing resources are devoted to ensuring that disabled people can participate fully in society. Legislation outlawing discrimination has been enacted in many countries and support groups are vocal in demanding full civil rights for those with disabilities. The idea that termination of pregnancy is going to significantly reduce the level of disability fails to take account of the fact that childhood onset disorders represent only a small proportion of the total number of disabled people in society. Most disability arises as a result of accident or through diseases of middle or old age. Termination of pregnancy for most of these is unlikely to gain widespread acceptability, particularly in light of the fact that most pregnancies, where testing is undertaken, are wanted pregnancies and the decision to terminate in such circumstances is not taken lightly. Even if screening for a wide range of inherited disorders becomes technically possible and economically attainable, large numbers of people will still slip through the net. Nor does it take account of the rate of spontaneous mutation, which can be high for some conditions (as much as 25% for Duchenne MD).

To withhold testing where the opportunity to provide information exists and where the patient wants it, because of a fear of possible long term consequences that are unproven seems to me to be unjustifiable.

Conclusion

Although genetic data is not without its risks. Many people see the availability of tests for an increasing range of conditions as a welcome benefit from scientific advance and a step on the road to the eventual development of cures.

Even without cures, the opportunity to be tested gives greater autonomy to individuals, whether or not they choose to avail themselves of the information.

Clearly genetic testing must be regulated in order that positive benefits can be accrued and abuse prevented. In framing regulations it must always be remembered that 'the perfect is the enemy of the good' and too Draconian a code, which eliminates any possibility of abuse, would probably also stifle innovation and leave families struggling in situations where a possible way forward already exists - and this must be avoided as a greater evil.

The European Alliance of Genetic Support Groups, Their Ethical Code and the Provision of Genetic Services

YSBRAND POORTMAN

The European Alliance of Genetic Support groups (EAGS) is a European umbrella organization of patients' organizations concerned with genetic disorders. It represents European umbrella groups for specific conditions and national umbrella bodies for the whole range of genetic disorders.

EAGS was founded in 1992 during a meeting in Elsinore (near Copenhagen) after a preliminary meeting in Leuven, Belgium (1991), and since meets every year, satellite to and in co-operation with the annual congress of the European Society of Human Genetics (ESHG).

42 associations from 14 countries are affiliated to EAGS. It is estimated that EAGS represents via its member organizations over 6 million registered families in Europe.

Objectives of EAGS

Main objectives of EAGS are:
- to serve the common aims of individuals and families with genetic disorders
- to promote the provision of services which meet the needs of these individuals and families.

Subsidiary objectives are:
- to facilitate the availability of reliable information concerning genetic disorders, their prevention and treatment
- to increase public awareness concerning genetic disorders
- to stimulate research into the causes, prevention, diagnosis and treatment of genetic disorders and to promote public understanding of the importance of such research
- to ensure the availability of comprehensive counseling on the options available and the consequences of particular choices
- to ensure the individual's right to autonomous and informed decision making

- to promote the assessment of the ethical, legal, psychological and social impact of the increasing understanding of human genetics
- to voice and promote the views of the members on issues of common concern
- to represent the members in European and International bodies and organizations
- to encourage the formation of new National and European umbrella organizations.

Bottlenecks

During the meeting in 1992 of the EAGS in Copenhagen bottlenecks were listed by the representatives of the various member organizations. The complaints and problems, regardless from what country the participants came from, were the same or very similar. These statements were recently reviewed and considered still to be up to date and relevant.

The bottlenecks were summarized as follows:
- primary and secondary level health care is not capable of dealing with genetics and genetic counseling; level of awareness of health care officials and caretakers about genetics is very low or absent
- relevant information reaches the people concerned (too) late or not at all
- appropriate and in time decision making is often endangered by late and inaccurate diagnosis, inefficient referrals, unbalanced and unreliable information
- research into the causes of congenital disorders is scarce and its importance is underestimated
- the impact of genetics and congenital disorders on the people concerned is underestimated; adequate and well trained guidance is hard to find
- scarce availability of reliable, well balanced and clear educational materials tailored to the various target groups such as health care workers, media, the public and the people concerned

Priorities

EAGS determined its priorities which read:
- public awareness (to make available information to the public in order to increase the general understanding about genetics and genetic disorders, in particular the new genetic technologies and their applications)
- education (of e.g. medical and general practitioners, health authorities and teachers) in matters relating to genetics, genetic disorders and genetic technologies

- encouragement of scientific research in (medical and community) genetics and of the free flow of information among scientists
- ethics (to promote ongoing assessment and research into the ethical and social impact of the new knowledge arising form genetic research)
- reproductive choice and screening (to promote access to comprehensive genetic counseling at genetic service centers, any decisions in this regard to be made on the basis of free informed choice of the individual or couple within the legal framework of each country. Such individuals or couples have the right to the understanding and full support of society for whatever decision is reached. They also have the right to complete information and confidentiality.)

Threats

There was also unanimous agreement about the concerns and the threats which were formulated as follows:
- scope and confidentiality of genetic testing and screening
- discrimination and stigmatization
- eugenic pressure
- commercial exploitation of human genome data
- society is not prepared for genetic progress
- equity of benefits of human genetic research.

Discussion of the concerns lead to the need for an ethical code. A working committee was installed which drafted a proposal - with the professional advice of the ethicist Mrs. Jeantine Lunshof. This proposal circulated among the member organizations. After many interesting discussions the sixth annual general meeting of the EAGS endorsed the proposal in London on April 12, 1996.

EAGS' Ethical Code

In the light of recent rapid scientific advances, EAGS has recognized the need for a basic set of statements pertaining to the ethical issues created by the increased understanding of human genetics.

EAGS endorses the standpoint agreed by the WHO:
„*Genetics and biomedical technology open up vast avenues for research and can provide humankind with much needed therapeutic tools. But, where human life and dignity are at stake, technology cannot be left to govern ethics on an empirical basis.*" (WHO, Summary statement on ethical issues in medical genetics. Geneva, February 1995).

EAGS, therefore, expresses its basic tenets in the following statements:

Statements and Guidelines

1. Medical genetics should serve the interests of the individuals who are affected or at risk from a genetic condition.
2. Needs-based access to information and facilities should be safeguarded.
3. Individuals should be free to decide for themselves whether or not to make use of the available information and facilities.
4. Genetic services are to facilitate diagnosis and provide options for and leading to informed decisions about preventative measures and/or treatment, and the consequent provision of appropriate needs-based services.
5. The continual improvement of the quality of life and of care and support of those affected by hereditary or congenital disorders is to be promoted, particularly by encouraging and supporting any necessary changes in legislation at national and at European level, to the benefit of patients with genetic disorders.
6. Persons with a disability or disease are entitled to unrestricted acceptance and solidarity from society.

EAGS calls on behalf of families with genetic and congenital disorders for:
- equal access to full information
- early diagnosis at accredited centers
- the maintenance of confidentiality
- the freedom of choice for all within the legal framework of each country.

Commentary to the Statements and Guidelines

1. Genetic counseling should be undertaken by qualified personnel. It should be comprehensive in its scope. Preferably it should be available in clinical genetics centers or in affiliation with such centers.

   Clinical genetics centers cannot be developed in every country immediately, but the demand for them derives its legitimization from the complexity of medical genetics as such and from the need to create expertise to interpret or manage this complexity.

   Genetic counseling should offer:
   - information about and knowledge of the hereditary aspects of conditions or diseases
   - referral and access to consultation with medical doctors, specialized in diagnosis and treatment of a given condition or disease.

   Options should be offered:

- for consultation with qualified social workers, psychologists and other relevant professionals regarding the social and familial consequences of the disease,
- for the meeting with representatives of patient organizations, if the counsellee wishes so.

2. Information about genetic services should be made available to all who may benefit from it. Genetic facilities should be within reach, both geographically and financially, for all who wish to make use of them.

3. Knowing one's own and/or one's partners genetic make-up creates options for action regarding genetic testing and reproductive choice. The decision concerning the preferred course of action to be taken in the light of this information rests solely with the individual or the couple, within national legal frameworks. There should be no third party coercion. This also applies to the option for prenatal diagnosis and the freedom to act upon the consequences.

   Utilization of genetic services must be voluntary. Any pressure to utilize all available technology for diagnosis or risk assessment should be avoided.

   The principle of privacy protection and respecting a wish not to be informed may interfere with moral obligations towards relatives at risk.

   Within the medical and institutional setting adequate measures for data protection should be safeguarded.

4. The general goal of genetic research is treatment of genetic conditions.

   Genetic services are to be directed towards accurate diagnosis and treatment of genetic conditions and to provide information concerning appropriate care and other options, if no effective treatment is possible.

   Genetic information can enable individuals to adopt strategies for reducing the risk of or preventing certain conditions, e.g. lifestyle changes, dietary measures or the avoidance of certain occupational hazards. This does not dismiss companies from their responsibilities for safer workplace and environmental conditions.

5. For disabled persons the limits to self-determination and the opportunities to live a fulfilled life are set by their living conditions and by the standard of care and support that is available. These depend largely upon the structure and organization of national facilities for health and social care, and as such are amenable to being influenced by public pressure or other factors - unlike the pattern of the disabling condition.

   Member organizations of EAGS regard the articulation of the needs of those affected by genetic disorders and to initiation of action that will bring services into being that will respond to these needs appropriately and effectively as a major task. This will involve action at national and at European level.

6. The existing and ongoing debate on priority setting in health care reveals the possible conflict of interests between the individual and society.

   Discrimination against disabled persons should be excluded. Persons with a disability or disease are under all circumstances entitled to full civil and human rights and to participate fully in all aspects of the society in which they live.

References

Selbsthilfegruppen und Humangenetiker im Dialog: Erwartungen und Befürchtungen (1993) Klaus Zerres & Reinhardt Rüdel (Hrsg) Ferdinand Enke Verlag, Stuttgart

The ethical aspects of biomedical research and the biomedical industry (1994) Netherland Haemophilia Society in cooperation with the European Haemophilia Consortium

Biomedical Research and Patenting: ethical, social and legal aspects (1996) European Platform for Patients' Organizations, Science and Industry

What Could a Balance Look Like between Individual Autonomy and Society's Need to Regulate?

Wolf-Michael Catenhusen

It was a good 15 years ago that the far-reaching options available for direct analysis of genes as the carriers of a human being's individual hereditary information were realized and became the subject of debate in scientific circles and in society in general. In 1983, the Presidential Commission set up by President Carter submitted an opinion in its „Splicing Life" report on the perspectives for genome analysis. In 1987, the German Bundestag's Study Commission on „Opportunities and Risks in Genetic Engineering" also submitted a report and recommendations on the prospect for genome analysis. Then available research findings, and including already familiar indirect analytic procedures, both reports made an attempt to develop standards for the responsible use of genetic tests at DNA level and for employing the individual genetic data produced by such tests. The 1987 report to the German Bundestag noted that *"direct genome analysis at DNA level is the most instructive analytical procedure for determining genetically related properties and for identifying the genetic causes of a disease or specific disposition„*. In previous work on detecting the genetic causes of diseases, the focus has been on analyzing a defect in a single gene responsible for one of the three to four thousand monogenetic hereditary diseases. Increasingly, however, we are having to identify disruptions to a gene that is largely responsible for causing the breakout of diseases based on a multitude of factors (breast cancer gene, cancer gene, rheumatism gene, Alzheimer gene). What may also be involved is detecting a deviation in a gene's structure from its "normal state", even where this does not enable us to make inferences about a person's current or future health. The „Human Genome Program" has augmented our knowledge of the sequences and functions of genes by leaps and bounds. This knowledge is crucial for an appreciation of biological processes and for clarifying the genetic context of important diseases. It offers new approaches for developing drugs and vaccines and creates grounds for hope. At the same time, a large number of genetic tests have found their way into medical practice, and the costs of individual genetic tests have been tumbling. Initial tests for the breast cancer gene can now be had on the free market for less than 300 dollars. How we handle such tests and use the genetic information gained in the process must be subject to clear regulations owing to the nature of such data. Both science and the general public must have a common interest in rules being drawn up to govern responsible genetic tests and the use of their results. The arrangements could consist of clear rules for the conduct of physicians, patients, test providers, employers and insurance companies. However, there may also be a need for such arrangements where individual autonomy, i.e. an individual's free decisions subject to his own responsibility, is to be created or safeguarded.

Emancipated citizens in our democracies have a right to as much knowledge as possible on the state of their health, using the latest diagnostic processes. Their right to informational self-determination also applies in principle to the use made of genetic tests. But autonomy must also include a right not to know. The basis for autonomous decision-making is greatly widened if precise, information is made available to interested parties on the quality and informational value of a test. In the name of autonomy, firms have now started selling tests checking a family disposition for breast cancer of the BRCA1 type, although there is no therapy available as yet for the condition diagnosed in this way. As long as physicians – whose professional ethics requires them to help people – are involved in implementing and evaluating genetic tests, the use of medical diagnostic services must be capable of being aligned to this goal. In my view there is no sign of this in today's commercialization of breast cancer tests.

Also, the right to informational self-determination does not automatically include the right to have all technically feasible health examinations using genetic tests financed by a sickness fund. In view of the limited funds available, cost-benefit considerations play a role. The benefit of genetic tests will be evident wherever they yield opportunities for prevention or therapy.

As I see it, the question of safeguarding decisions subject to personal responsibility also applies as long as possible to the sensible area of prenatal diagnostics where, since the end of the 1980s, increasing use has been made of genetic tests at DNA level. Prenatal diagnostics concerns the detection of serious incurable or hereditary diseases. If a mother or parents decide against the continuation of a pregnancy and have an abortion, they are not liable to punishment. What we also have to consider, however, is that it has become possible to diagnose monogenetic hereditary illnesses prior to birth using genetic tests which will enable the child to live with a quality of life for decades. So, in future we will be dealing with risk factors whose relevance for the fate of a child cannot be precisely established. With the increased use of genetic tests, we will, in the long run, be abandoning the previous goal of prenatal checks, viz that of exploring the suspicion of a serious, incurable disease in a child prior to or after birth. If this linkage appears in medical practice, however, I fear that decisions will be made against a child, decisions involving eugenic objectives. With a view to preventing such undesirable developments in the name of freedom of decision, we will go on needing an interplay of genetic and social advice. Advice given must be unbiased and given to enable those involved to come to an informed decision. To this end, we need to provide relief for those living with handicapped children; **and** we need parents' right to decide against prenatal examinations of their children without this giving rise to a refusal by the welfare state to continue its solidarity payments. Genetic data must be protected against access by third parties. Any genetic data obtained by a physician must remain with him and be subject to his professional secrecy.

Genetic tests not only help open up new opportunities for prevention therapy. The results may also be of interest to third parties. Passing them on may entail serious risks of genetic discrimination. Parallel to the introduction of genetic tests as standard checks, therefore, we need decisions by society on the extent to which personal data gained from genetic tests must be protected against access by third parties.

1. In many countries the use of genetic fingerprinting to identify perpetrators is subject to regulation. The aim of the statutory arrangements in Germany is to confine the use of this procedure to people accused of a serious crime, and to rule out any attempt to obtain information on a defendant's personality features using genetic tests.

2. The United States and some Western Europe countries have a moratorium or specific draft legislation designed to rule out genetic testing as a prerequisite for taking out health or life insurance.

3. Especially in the mid-1980s, the use of genetic tests at the workplace was regarded as particularly problematical. Technical developments have not yet confirmed the concerns aired at that time. So far, genetic tests at DNA level have not proved to be so interesting for occupational medicine as many experts had anticipated at the beginning of the 1980s. One minimal demand discussed at present concerns ruling out DNA testing in pre-employment medical checks.

At the present level of development in genetic testing, I think, it is possible to obtain a balance between individual autonomy and society's need to regulate. However, this state of affairs may well change in the near future. Various firms in the US are developing processes for testing DNA samples of a number of patients at low cost – possibly even without help from a physician. It will then be possible to check a sample for hundreds of potential mutations.

Moreover, genetic tests in the context of a changing health policy could acquire a different quality. In the feasible concept of predictive medicine, serial examinations would pinpointedly detect the carriers of particular risk factors: with knowledge of the problem those concerned could then be given an opportunity to take precautionary measures, but also to live a healthier life or, by opting for an abortion, to prevent their own risk factor being passed on to their children. In such a context, an individual's autonomy might be restricted by societies higher-ranking expectations.

New Genetics, New Ethics?

Matthias Kettner

Introduction:
Three Standard Responses to Moral Pluralism

The new genetics, like every groundbreaking innovative technology with a potentially global scope of application before, challenges established social habits and disturbs ways of thinking that seemed to be settled.[1] What is the place of ethics in the wide but uncertain set of responses that are called for if we want to shape and normatively govern the emergent genetic technologies rather than be governed by them as a matter of fact? Of the fringe benefits generated by the Human Genome Project for concern with ethical, legal and social issues, a good deal is spent on the presumption that there *is* such a place for ethics. This presumption is typically expressed in statements like the following taken from the *Encyclopedia of Bioethics:* „The explosive development of new knowledge and techniques in medicine and biology has made bioethics one of the central areas of practical, moral concern. And those seeking to solve moral problems in this area naturally appeal to philosophical ethics" (p.726). Hence biomedically applied ethics should have a say. In the copious literature about the Genome Project's „ethical issues" one frequently encounters the much stronger presumption, overtly expressed or covert, that ethics should be in the driver's seat with regard to the attempted normative governance of the micromolecular genetic revolution in medicine and other practical domains.

The strong presumption gives pride of place to ethics. Naturally, this view does not go unchallenged. A case can be made that moral thinking, for all its worth as a resource of reflective governance, has but a very slight bearing on the social shaping of big science and technology. The applied-ethics literature devoted to matters genetic is strangely mute on self-critical comparisons between the tangible guid-

---

[1] Following Wertz & Fletcher (1990), Wheatherall (1985), and Comings (1980), I use the term 'new genetics' as an umbrella notion covering all uses in the field of medical genetics of technologies for *direct* DNA analysis or manipulation in distinction to the „old" technologies for the analysis or manipulation of DNA *expressions.*

ing function that „applied ethics" is supposed to have, and the real impact of applied ethics on the emerging practices of applied genetics, for instance on genetic
services in Europe (Harris et al. 1997). Both the utopian and its complement, the
cynical attitude, are inherent temptations in the practice of applied ethics, when
'applied ethics' is properly understood as covering all endeavors to utilize philo-
sophically more or less refined resources of moral thinking in order to make
things go morally better than they otherwise would, within targeted particular
realms of social practices (Kettner 1992).

The Berlin conference that prompted the following reflections evidently endorsed the weaker presumption.[2] Fortunately, it was not committed to the stronger one. Unfortunately though, it dodged two simple but very consequential questions:

(Q1) What kind of ethics should have a say in the reflective governance of the new
genetics, given the fact of moral diversity both on the level of philosophical theory
as well as in the political arena of such states, predominantly western stile liberal
democracies, where bioethics has gained more political momentum than an academic fad?

(Q2) Given that a wealth of issues are attributed to the new genetics and are perceived to be somehow important, usually in amplification through the mass media
(Kettner & Schäfer 1997a), how are we, or other protagonists trying to apply ethics
to the new genetics, to sort moral from non-moral issues?

Q1 invites the response that perhaps *all* kinds of moral beliefs should have a say in
attempts at reflective governance of the genetic technology, especially in the medical context, since patients (not to speak of doctors) bring to the medical encounter
all kinds of beliefs which they take as moral in import and as relevant for the ensu-
ing medical transactions. I think this quick response draws on a sound intuition.
However, Q1 now turns into a question that evidently stands much in need of
more theoretical elaboration, namely, what kind of ethics can do justice to moral
pluralism, that decidedly modern condition of application, without surrendering
the normative intention of applied ethics (i.e. the intention to set forth as valid,
and to vindicate the corresponding claims of, standards for the assessment of
judgments in particular moral matters) to a parochial relativism of one's closest
cultural peer-group?

---

[2] A preliminary version of this paper was presented 1997 in Berlin at workshop on „The New Genetics - From Research into Health Care. Implications for Users and Providers". I would like to
thank the participants in the discussion and especially Irmgard Nippert for broadening my appreciation of the emerging field of reprogenetics.

Three interesting responses are salient in the literature on bioethics. The first typical response is exemplified by Max Charlesworth's important book on „Bioethics in a Liberal Society". Applying ethics in a liberal society, according to Charlesworth, boils down to person-relative perceptions of moral choice, with a thin overcoat of transpersonally shared and legally reinforced moral norms, a moral norm that one ought to let be others' moral convictions even when one judges them improper or wrong, and a second transpersonal moral norm limiting the first one, namely that the state (or some other effective organ of a body politic) is morally responsible for politically regulating the exercise of free individual choice such that no-one (in one's own political society) suffers serious harm from anyone's exercise of free individual choice.[3] This position embraces a normative relativism: it holds that it is ethically wrong to pass ethical judgment on the behavior and practices of another individual, group or society with a substantially different ethical code, or that it is wrong to intervene into the affairs of another individual, group or society on the basis of such ethical judgments. A difficulty with this otherwise attractive position is that it is either incomplete (since it requires a consensus on liberal core-values which it prescribes in a supposedly non person-relative way) or inconsistent (if such a consensus is required but permitted to be person-relative) or arbitrary (if a consensus on liberal core-values is viewed as the contingent outcome in only some lucky few societies in world-history).

A second typical response to Q1 consists in the establishment of a small set of abstract moral „principles", to secure a practical use for these principles as facilitating tools for ethical decision-making, and to diffuse these instruments (mostly through channels of professional education) as wide as possible through the medical community. Consequently, more and more people will fall back in their moral thinking on what for them has become a salient interpretative authority, for instance , the „four principles of bioethics" (Beauchamp & Childress 1989) associated with the famous Kennedy Institute of Ethics. There is much to be said in favor of a principles-based approach to normative thinking about bioethically interesting issues. For instance, the approach provides a background of normative reference against which it becomes easier to realize and articulate disagreement. Perhaps the approach furnishes what might be called with a linguistic analogy the „Esperanto" of bioethics. However, there is also a lot to be said against overreliance on a small set of abstract moral principles. Clouser (1996) and others have stringently criticized this way of doing bioethics concerning concrete target

---

[3] „The liberal society is characterized by unconditional respect, as Kant would say, for personal autonomy, and that carries with it a respect for ethical pluralism and a resistance to the state and the law intervening in the realm of personal morality or ethics" (Charlesworth 1993, p.153). „In a liberal society people should as far as possible be allowed to make their own moral decisions for themselves and it is not the business of the law to enforce a common code of morality. The law should be brought in, so to speak, only when other people are likely to be harmed in some obvious way" (ibid., p.74).

practices such as genetic counseling.[4] Lacking an underlying justificatory connection between them, the elements of a small set of abstract moral principles that have won salience within a large community of bioethicists, such as respect for autonomy, nonmaleficence, beneficence, and justice, function merely as check-lists that name issues worth remembering when one is considering a biomedical moral issue. Paradoxically, in the hands of the experts who are translating such principles into practice and who monopolize the interpretative power that is required for doing so, those principles tend to open a backdoor for the very paternalism in moral matters that they manifestly purport to defend against.

A third response is represented by Bernard Gert's work on the moral system of common morality. Gert makes a convincing case that a number of „moral rules"[5] are accessible in the background (if not on the forefront) of virtually every normally socialized reasonable adult person. „Common morality does not provide a unique solution to every moral problem, but it always provides a way of distinguishing between morally acceptable answers and morally unacceptable answers; that is, it places significant limits on legitimate moral disagreement" (Gert 1996, p.29). Common morality is tied to our most fundamental normative distinction, the distinction of rational versus irrational courses of action, in that support for the public system of moral rules is grounded in the fact that general observance of this system adds to others' actions being reliably not irrational. Rationality and morality have a common denominator in the avoidance of harm.

The normative authority of reference to moral rules which belong to a rationally defensible layer of common morality finds its limits in the fact that concrete interpretations and relative rankings of commonly recognized harms (death, pain, disability, loss of freedom, loss of pleasure) differ across different moral points of view. The rules themselves do not specify any rational procedure for handling in morally just ways the corresponding space of rationally underdetermined differences. Yet the capacity for handling such differences in morally just ways can be seen to constitute a bench-mark for the „application" of ethics, such as bioethics, in culturally pluralist societies. Differences between moral points of view can run as deep as to the very idea of what constitutes specifically a *moral* aspect in distinction from a non-moral aspect in dealing with a concrete problematic issue.

Here, Q1 (the question, What kind of ethics should have a say in the reflective governance of the new genetics, given the fact of moral diversity?) is linked with Q2 (how are we to sort moral from non-moral issues in applying ethics?). It is therefore necessary to advance the normative theory of applied ethics in two com-

---

[4]   See Gert (1996a, 97–124) for a discussion of a principles-based approach to nine Huntington disease
      cases in comparison with the rationally reconstructed common morality preferred by Gert (1998).
[5]   Gert (1988; 1998) has articulated ten such rules (for instance, „do not kill!", „do not disable!", „do not
      deprive of freedom!", „do not deprive of pleasure!", „do not deceive!") and has given ample proof of
      their relevance with regard to issues in bioethics (Gert 1996; 1997).

plementary directions: not only with a view to the nature and justification of common morality, as Gerd does, but also with a view to grounding a rational procedure for the handling of, and moral judgment across, moral diversity both with reference to discordant moral judgments about the same issue within a common frame of reference (e.g. moral disagreement within the framework of the four principles of bioethics about the moral permissibility of human cloning) and also where moral pluralism is the source for disagreement, intolerance, or indifference. This desideratum can be realized with a consensualist framework of applied ethics - call it *discourse ethics* - that I am going to outline in the remainder of this paper (sections II, III, IV). At suitable points in my exposition of discourse ethics I will look at some of the many morally sensitive issues of the new genetics that have surfaced in the course of our conference (sections V, VI).

Moral Horizons, Discourse Ethics, and Other Moral Paradigms

A moral point of view or horizon (as I would prefer to say) is the evaluative stance of someone who identifies with some morality M (e.g., by „having been brought up" in the spirit of M, by finding M convincing, etc.) for assessing practical activities and the reasons for which people take themselves to be justified in doing what they do. A moral horizon discloses the „moral costs" (according to the particular value standards that are acknowledged in M) which accrue to people's transactions. By considering something in a moral horizon one assesses impacts of, and reasons for, actions with an eye to minimizing what moral costs it may have. A moral horizon, being a *moral* point of view, discloses what is right or wrong (as judged by its avoidable moral costs) and what thus ought to be done or ought not to be done (on moral grounds). Being a *point* of view, no moral horizon is all-encompassing. Being a point of *view*, any moral horizon, to the extent that it can be called rational, must admit of controversy and consensus, of questions and answers, of argument and counter-argument.

Discourse ethics is an emerging new paradigm of normative philosophical moral theory. Roughly, four paradigms of normative moral theory and their respective central principles are presently in the foreground of philosophical ethics. These paradigms are the following. (1) Kantian deontologism with its principle that persons ought to be respected as ends-in-themselves, (2) utilitarianism with its principle that utility ought to be impartially maximized, (3) contractualism with its principle that explicit or tacit agreements for mutual benefits ought to be honored, and (4), consensualism with its principle that all normative arrangements ought to be procedurally governed through free and open argumentative dialogue („discourse"), ideally of everyone concerned (Apel 1980 and 1989; Habermas 1993).

Of these, consensualism as developed into a „Communicative Ethics" or „Discourse Ethics" (Apel, Habermas) is the most integrative and most flexible position. The initial stimulus behind the philosophical development of discourse ethics is the intuition that the reasons on which people claim that something is morally right must be such as to be conceivably acceptable from the first-person plural perspective („we") of everyone concerned by the practice, activity or regulation whose moral rightness is at stake. Call this the *discourse principle*. Moral rightness then is a property of action-norms, a property ultimately dependent on the cooperative discursive practice of free and open dialogue between rational evaluators about discordant appreciations of allegedly morally sound reasons.

Of course, not all moral content is held to be *generated* in dialogue. Nor are we to devote all our moral life to argumentation. Rather, the discourse principle is a problem-driven principle i.e. its critical force is invoked only when particular issues cannot, or can no longer, be satisfactorily handled by the conventional resources which the people concerned are used to take for granted in the respective practical contexts. The discourse principle operates on subject matters that are always already pre-interpreted by whatever moral intuitions the participants happen to bring to the fore.

Applying discourse ethics means transforming, rather than substituting, the already extant moral infrastructures of targeted domains of social practices. Moral discourse is the medium to modify and reshape them. In moral discourse, people work through their various moral perplexities in a cooperative effort at reaching a maximally value-respecting practical deliberation which everyone can support - though such discursively prompted deliberation need not totally coincide with what each claimant would personally judge as the right way to go, given only each claimant's own moral horizon and supposing that other moral horizons were not part of the problem at hand. In fact, it may deviate considerably from „the" right exclusively within ones own moral horizon. However, there is the possibility of integrity preserving genuinely moral compromise (Benjamin 1990) reached through procedurally fair negotiation (van Es 1993) *within* a moral discourse.

A second point deserves mentioning. Consensus building constrained by discourse ethics allows to emulate central principles of other moralities. For instance, if all people whose needs and interests are affected by some practice p were to agree in a practical discourse that p should be regulated by, say, utilitarian standards then the discursively prompted consensus about the morally right way of regulating p will result in p's being regulated so. Yet whatever substantial moral principle people would want to adopt (e.g., a utilitarian principle of maximizing the average satisfaction of individuals' preferences) will become constrained in discourse ethics by respect for the capacity of people to reach a common understanding about how they want to treat and be treated by others, regardless of egocentric positional differences.

Five Parameters of a Moral Discourse

Discourse ethics grounds a number of constraints on consensus-building about normative matters, including consensus-building about dissensus (Apel 1987; 1993; Rehg 1994). These constraints, if jointly and sufficiently fulfilled, bestow rational moral authority on the corresponding consensus-building process, whereas failure to sufficiently satisfy one or more of these constraints weakens (perhaps to the point of invalidating) claims to moral authority for the prompted consensus. Discourse ethics, in the new perspective I propose, which has outgrown Apelian and Habermasian formulations of discourse ethics, vindicates a moral resource of very minimal morally normative content but with the advantage of being ubiquitously available, namely an ethics that resides in the normative texture of the very practice of argumentative discourse. In the order of justification, then, discourse ethics starts by making this minimal morally normative content explicit - by a heuristics of what is technically called the avoidance of „performative self-contradiction" (Apel 1987). From this unassailably grounded moral resource - call it: an ethics *in* discourse - discourse ethics then seeks to develop the much richer notion of a fully rational moral discussion (a discussion fuelled by discordant moral judgments on some morally perplexing matter) - call this the notion of a *moral discourse*.

The notion of a specifically moral rational discourse is perhaps the most important theoretical contribution of discourse ethics to applied ethics. Moral discourse is rational argumentation about problems that are perceived specifically as *morally* perplexing, i.e. problems that are perceived to call for answers in terms of deontic judgments (expressing what we think may or ought or ought not to be done) which are underwritten, not by any sort of reasons, but specifically by reasons that we recognize as *moral* reasons (thus rendering such deontic judgments *morally* deontic judgements). The qualification that moral discourse be rational is but a corollary of an aspiration to argue about conflicting moral reasons with a view to (re)making certain morally deontic judgments and their reasons more coherent where they threaten to come apart in the face of disturbing moral perplexities.

To bring out more clearly the sense in which moral reasons, judgments and norms differ from non-moral reasons, judgments, and norms, it is helpful to consider the notion of *moral responsibility*. Moral agents are persons who bear, and take their moral peers to bear, moral responsibility. An agent's moral responsibility is a responsibility that is neither exhausted by that agent's *causal* role in the outcome of actions nor in that agent's liability that is relevant to a *juridical* assessment. Rather, the bearing of moral responsibility consists specifically in collectively taking seriously, with a view to minimizing moral wrong [=moral costs], how the outcome of one's conduct, i.e. of possible actions or omissions, affects oneself or relevant others and their relationships for good or ill. Moral reasons are distinct from other sorts of reasons in that they link someone's judgment to some community's generally acknowledged specific interpretations of moral responsibility.

Different moralities, of course, assign different contents to the structure of such responsibility: different ways of acting („conduct"), different reference groups („others"), different significant values („good or ill"). This diversity of interpretations implies moral pluralism. From pluralism we have to distinguish moral relativism. Diversity and pluralism do not at all imply relativism. Consider: Like natural languages, moralities are a pervasive feature of human culture. Yet whereas translatability between any two natural languages seems to be feasible in principle without any limitation, the fact of moral diversity (and value pluralism) has been taken to support the view that it makes sense to speak of justified moral claims or commitments only with reference to particular cultures: cultural moral relativism. However, strong moral relativism (and incommensurability of values across deeply different cultures or across subcultures) is a self-refuting theoretical view, as is the view that understanding across different value-horizons is impossible. I cannot discuss these points here. Suffice it to say that the world-wide recognition specifically of those deontic reasons and morally significant values that we abbreviate as *human rights* in fact contradicts any strong relativism and incommmensurabilism. At a sufficiently abstract level (where moral deontic reasons take the form of „moral principles") there is considerable overlapping consensus across diverse cultures despite disagreement over their more fine-grained interpretation, their ranking, and to what extent which real social practices ought to be governed by which principles (Outka & Reeders 1993). A certain range of values (e.g. freedom, the provision of basic needs, integrity of primary affective bonds, sanity) bear moral significance virtually everywhere (Gert 1998). Yet their determinate interpretations in terms of moral action requirements, moral norms, morally deontic judgments may vary dramatically across substantially different cultures.

The fact of moral diversity needs to be sensibly accommodated in any „rational" morality and normative moral theory with universalistic aspirations. To see why, consider that universalistic moral claims make a claim on everyone who properly takes them into account intellectually. They purport to command the assent of whoever is a morally responsible rational agent. Yet the people making such claims are always members of some particular community in space, time, and culture. Hence they run the risk of imposing the claims of what they take to be their universalistic morality on others whose moral views, if only they were allowed to express themselves, could be seen to differ from or even to defy the imposed claims. Owing to the universalism inherent in our notion of rational validity, any markedly „rational" morality will also aspire to universalism. Yet a sensible universalism in morality, it seems, must not impose rigid moral principles on an unruly moral world of heterogeneous moral views. Were it to do so, it would be buying uniformity at the cost of dogmatism or paternalism. Both dogmatism and paternalism concerning the imposition of alien moral views spell unnecessary „moral costs". They are therefore wrong according to the very standards of a truly rational morality. To the extent that an allegedly rational morality or moral theory is insensitive to its own impact, or lacks the conceptual resources for the moral assessment of such impact in its application, it is seriously inadequate to the modern pluralist condition.

Whenever in some target domain a reflective mode of governance, like argumentation, can be brought to bear on normative change (as applied ethics presumes), then the corresponding processes of argumentation represent moral discourse if they embody and express a set of parameters which jointly guarantee the moral integrity of the discursive power that is exercised by the respective community of argumentation.

I will merely list five normative parameters that are characteristic conceptual elements in the idea of a moral discourse. I have given a detailed vindication of these parameters elsewhere (Kettner 1998c). Briefly, the justificatory route on which one arrives at these parameters starts out from an uncontroversial notion of argumentation (being the rational evaluation of seemingly good reasons with other apparently good reasons, cf. Kettner 1999) and from a realistic view of norms (as interlocking socio-psychological patterns of appropriate ways of proceeding, cf. Kettner 1998b). Each of the following five parameters can be introduced as a well-grounded partial answer to an unavoidable general question, namely 'Are there recognizable proprieties such that if they were not mutually required among co-subjects of argumentation then argumentation in the face of conflicting reasons, specifically those representing interpretations of moral responsibility, would not make sense for co-subjects of argumentation?'. The parameters, then, answer to the goal of pushing rational argumentation specifically about moral reasons in the face of incoherent moral judgments as far as it goes.

Parameter 1: Reasonable Articulation of Need-Claims:
All participants in a discourse should be capable of reasonably articulating rationally any need-claim they take to be morally significant.

Parameter 2: Bracketing of Power Differentials:
Differences in (all sorts of) power which exist between participants (both within and outside of argumentation) should not be any participant's good reason in discourse for endorsing any moral judgment.

Parameter 3: Nonstrategic Transparency:
All participants should be able to convey their articulations of morally significant need-claims truthfully, without strategical reservations.

Parameter 4: Fusion of Moral Horizons:
All participants should be able to sufficiently understand articulated need-claims in the corresponding moral horizons of whoever articulates them.

Parameter 5: Comprehensive Inclusion:
Participants should make the following a constraint on what their community of discourse can accept as good reasons: that participants must anticipate whether their reasons can be rehearsed by all nonparticipant others who figure specifically in the content of any moral judgment determined by the participants to be taken seriously by everyone.[6]

Note that discourse ethical consensus-building is not equivalent to a unanimity requirement, nor to majority vote, nor to any preference-aggregative decision procedure (e.g. bargaining). The dynamics of consensus-building in practical discourse does not guarantee a unique „solution" to all moral issues. Staking out a *range* of permission is often the best we can come up with. No morality is an algorithm for solving problem cases. To some extent, morality must countenance tragic choices and persistent tensions. Such choices and tensions at best admit of alleviation, not of total resolution when considerable „moral costs" are bound to remain. However, a consensus sufficiently reflective of the parameters of moral discourse guarantees to all parties who mutually recognize one another as having a credible stake in the outcome of the discourse that they are mutually aware of all their different „moral costs", and that they are also mutually aware of the right-making reasons from every participant's moral horizon. Realistically, no rational morality can guarantee anything stronger than that. The possibility of reasonable *dis*agreement (dissent) exists alongside the possibility of reasonable agreement (consensus), notwithstanding the conceptual truth that the latter envelopes the former.

On this basis, a morally-discursively prompted consensus may well integrate some amount of justified dissensus. Depending on whether such dissensus expresses mutual and omnilaterally justified concessions, a consensus that is reached via moral discourse may as such express a moral compromise. In such a compromise, however, no-one's morally significant need-claims will have been compromised intolerably.

Summing up this section: Discourse ethics is a two-tiered normative moral theory. On the first tier, completely general yet morally significant norms of argumentation are identified, yielding a minimal morality („ethics in discourse")

---

[6] Space does not permit to comment the parameters. Their point, however, should be intelligible from their formulation, perhaps with the exception of the somewhat difficult to formulate but intuitively quite simple fifth parameter, its idea being that the role of a participant *subject* be available in discourse to anyone who figures in the role of an *object* in the content of any deontic judgments which the community of argumentation comes to endorse by way of that discourse.

whose claims range over, and whose grounds can be ascertained by, all subjects of argumentation. On the second tier, moralities are meta-ethically characterized as variations of a common basic structure of moral responsibility. Moral reasons represent how moral communities fill out this basic structure with determinate content. By tracing normative requirements that are arguably necessary for argumentation about moral reasons to retain its rational point in the face of moral perplexity and moral pluralism, a set of five parameters is proposed which together define (as a normative ideal type) the notion of a moral discourse. Moral discourses are reflective modes of governance. If governed by moral discourse, normative textures in transition will not deteriorate and may even progress in their moral qualifications. Moral discourse as specified by discourse ethics is a medium in which our moral convictions can face the tribunal of historical experience in divergent moral horizons.

Disputes about Facts, Values, and Norms in a Moral Discourse

Naturally, there are many different evaluative perspectives, only some of which are moral horizons. We can assess actions and the reasons for which they are performed from the prudential, the technical, the legal, the moral, the economic, the religious and, no doubt, from many other points of view. For instance, some new medical technique may be ingenious and highly recommendable on prudential grounds when seen from a technical point of view (e.g., the Triple-Test for fetal anomalies) and yet be dubious on grounds of its moral costs when seen from a certain moral point of view (e.g., because routinizing the Triple-Test for fetal anomalies socially reinforces unjust attitudes towards people with certain handicaps or their parents). A new law might be legitimate from a juridical point of view (e.g., the Chinese law on family planning) and yet be illegitimate when considered morally (e.g., because it violates human rights). A new entrepreneurial institutional arrangement (e.g., managed care in the United States) for all the sense it may make economically, may be found wanting medical-ethically when the patterned distribution of certain risks for certain patients (not to get the treatment that would be best for them) and certain benefits engendered by that instrument (e.g., savings in redistributable resources) is assessed.

There are no moral problems *per se*, i.e. independent of people who are morally perplexed by what their taking a moral point of view discloses to them about some of their practices. As substantial interpretations of moral responsibility differ, what is a moral problem to one person or group is not always a moral problem to another person or group, though both parties view things in *moral* horizons respectively. We find moral problems when we find people in doubt about whether a course of action is right or wrong. Hence if we want to understand moral problems, we must find out the rationale why people are perplexed about what is right or wrong.

Disputing what people take as proper responses to their moral perplexity, to the extent that it is rational, is governed by a logic of discourse. This logic of discourse revolves around our powers of raising and answering broadly distinct types of questions:

- Questions of fact (in any sense in which something to believe can be the case or fail to be the case).
- Questions of value (in any sense in which something to be appreciated can be good or bad).
- Questions of norms (in any sense in which something to do can be required of, or permitted to, someone by someone).

Questions of *fact* and their associated truth-claims can be disputed with reference to the availability and convincingness of the evidence for establishing what is the case. Questions of *value* and their associated claims of evaluative commitment can be disputed with reference to the appropriateness and importance of the properties in virtue of which something is held to be valuable, in the disputed sense of good. Whether the purported good-making or value-giving properties are really present is then again governed by questions of fact. Questions of *norms* can be disputed with reference to the values a norm is held to subserve or express. Whether the values in virtue of which it is claimed that certain agents ought to do certain things really authorize the norm in the disputed sense of requiredness or permittedness is then again governed by value questions and again, by factual questions.[7]

Two people (or parties) in disagreement about what one ought to do must consider whatever other norms they subscribe to and link up with the norm in question. Norms face the tribunal of discourse and experience corporately: commitment to some component normative texture N may turn out to mean, on pain of incoherence, subscription to (or refusal to accept) some other component normative texture N'. Furthermore, people turn to what each of them takes as the relevant values that bear on the norms. And two parties in disagreement about the sense in which they have reason to take something to be good must be prepared to be led into scrutinizing as many of their other evaluations as are found to be somehow related to the one in question. Values, like norms, face the tribunal of discourse and experience corporately, hence someone cherishing some value V may

---

[7] Using a philosophical technical term, one can say that normative differences *supervene* on evaluational differences which in turn supervene on factual differences. 'Supervenience' here is a conceptual relation such that if properties of kind x supervene on properties of kind y then there can be no difference in x without some relevant difference in y. For the concept of supervenience, see Kim (1994); for a pragmatic concept of evaluation and the handling of evaluational differences, see Dewey (1939) and Taylor (1961).

find himself committed, on pain of incoherence, to some other value V'. Further-more, people discuss what they take to be the relevant facts and their relations on which they think depends the sense in which they suppose something that is in question to be good. The unfolding dialogical dynamics of relating factual, evaluative and normative questions, if need be in many repetitions, make for ra-tional inquiry in the perplexity-driven discourse processes.

Morally Perplexing Factual, Valuational, and Normative Questions in the New Genetics

I will first illustrate the moral relevance of questions of fact with an issue that is very common in genetic counseling: Risk-Communication and the cognitive com-mand of probabilistic thinking.

(1) Against the background of an extended debate about the normative nature of counseling, specifically of genetic counseling, it is safe to conclude that the most viable conception of genetic counseling is that of a fiduciary dialogical encounter in which one party (the counselee, or client) seeks professional help in a situation of momentous uncertainty (usually a problematic situation that calls for a resolu-tion or personal decision to be made by the counselee) and in which the other party (the counselor) takes over a professional responsibility of exercising her professional skills with the primary goal that the client make a fully informed deci-sion, one that is based upon the client's own values and full information about the client's options and nothing else (Singer 1996). Counselors may legitimately pre-sume that their clients expect them to help them on with *valid decision-making*. „Valid decision making requires that the client be provided with adequate infor-mation, not be coerced by any member of the health care team, and be competent to understand and appreciate the information given. (...) The overall ideal of the counselor is to prevent a client from making an irrational decision, not merely one that would be irrational for anyone, but one that would be irrational for those hav-ing the values that the client has" (ibid., p.129).

It is often quite problematic for the clients when they have to reach a yes-or-no decision about matters that are heavily loaded with the opaque consequentiality of personal choices, in contexts of reproductive decisions often psychologically am-biguous choices, so that binary choice seems strangely inadequate. (Prenatal diag-nosis - yes or no? Acceptance of a handicap - yes or no? Abortion - yes or no?) The decision to be reached by the client can be construed as a singular deontic judg-ment, i.e., as the client's own judgment that she herself may (or may not), or that she ought (or not), do this rather than that, all relevant things considered. The subject matter of genetic counseling being what it is, for clients to arrive at their judgments they will in the majority of cases have to incorporate probabilistic information into their practical deliberations at some point. Such information is notoriously hard to understand in its true cognitive significance (Cheuvront et al. 1998; Shilo & Sagi 1989; Scholz & Kroner 1988; Lippmann-Hand & Fraser 1979).

For instance, in a counseling session with a woman who is upset about her prospects of developing breast cancer it may become necessary to cognitively integrate, at the level of questions of fact, the product of three different probabilities each of which comes with its own test-dependent margins of statistic uncertainty: The probability of carrying the etiologically relevant genetic modification, the probability of developing the corresponding malady, and the probability for a strong expression of the malady.

For a dramatic illustration of the cognitive demands that are involved in the appropriate handling of probabilistic matters of fact one only needs to ask people to elaborate a bit on their personal understanding of common sense probabilistic information, say, the weather report's declaring a „30 per cent chance of rain today". Depending on the clients' grasp of what is really involved in the probabilistic facts that are constraining their contexts of choice, their deliberations will have different possible outcomes, and the rational credentials of the particular deontic judgments for which they will finally settle will vary accordingly.

Note that I am not saying that different perceptions of what the facts are *determine* different normative judgments. To say this would be to ignore the decisive mediating role of values and of differences in value commitments. Think of two identical twins who learn (and correctly grasp the cognitive significance of the fact) that they are almost certain to develop Alzheimers disease within the next 10 years. If one of them continues his way of life as before while the other thinks he ought to make substantial changes in his long-term life-projects this difference in the normative result of their respective practical deliberations need not indicate arbitrariness or insufficiently rational deliberative processes. It may well indicate different value commitments: For instance, one of them might be strongly religious such that early death through genetic malady does not count as much for him as it does to his secular-minded brother.

(2) Let me now look at another morally perplexing feature of counseling that has received some attention during this conference: autonomy.

Historically, in the guiding ideology towards the client of the relatively young profession of genetic counselors one can read off a slow change from directiveness to non-directiveness. More recently, one can observe yet another far-reaching change in the profession's self-understanding, namely from a construal of the norm of directiveness and the norm of nondirectiveness as total opposites towards a new construal of this distinction that is yet struggling for its articulation but should at any rate sublate the former simple opposition.

Not only is the simple construal of directiveness and nondirectiveness as an all-or-nothing affair of polar opposites empirically untenable. It is untenable by common ethical standards as well: The ideal of the counselor as helping the client make valid decisions does not categorically proscribe directive behavior. Even confronting a client with what the counselor perceives as certain contradictions, indifferent feelings, or selective ignorance, is morally permitted or may even be morally required (under circumstances that can be detailed by reference to the five parameters of a moral discourse and typical counseling interactions) of an experienced counselor - morally permitted or required, that is, not in spite of but rather as an expression of respect for the client's autonomy of personal choice! A reconsideration of current notions of autonomy (e.g., current in the literature on „informed consent") seems much needed. Also, a reconsideration is overdue of the discrepancy that is found to exist between counselors' generally very high regard for „nondirectiveness",[8] and the not nearly so high regard for „nondirectiveness" expressed by their clients.

Not some ill-understood notion of „respect for autonomy" and „nondirectiveness", the latter often wrongly equated with the counselor's expression-in-action of the former,[9] should serve as the focal point in a refined normative understanding of genetic counseling. Rather, what should so serve are clients' needs to prepare themselves for choices such that the choices eventually made will tend to be valid ones rather than invalid ones for the clients. Standard analyses of „autonomy" and „non-directiveness" suggest that the autonomy of a client's choice depends essentially on that choice's being authentic, informed and non-coerced. In this format, social others (e.g., the counselor) as assistants of such choice are stylized into providers merely of whatever it takes the client to be *informed*. Beyond that, their role is not a positive one (as, e.g., a facilitator or contributor) but merely a negative one, a work of refraining and keeping themselves out of their clients' „personal choices". However, the standard view of autonomy and non-directiveness is ill-fated. As the hegemonial ethos in the context of genetic counseling, it is neither empirically sustainable (Bartels et al. 1997; Michie et al. 1997; Wertz & Fletcher 1990) nor morally justifiable (Walters 1993; Caplan 1993).[10] The new genetics should provide a strong and highly specific stimulus for the on-going reconsideration and critique of received notions of autonomy in applied ethics (Jennings 1999).

---

[8] Wertz & Fletcher (1990, p.459) found that „nondirectiveness is the most widely ingrained approach to genetic counseling, when viewed cross-culturally".

[9] Bosk (1993) referring to the 70s and 80s speaks of the „workplace ideology" of genetic counselors.

[10] Wertz and Fletcher, surprised by finding „impressive cross-cultural differences of opinion" (p. 77) both about good counseling practice and moral points of view, conclude that these findings tend „to refute the view that the diffusion of technology carries a Western cultural tradition that resolves ethical disputes in the name of patient autonomy" (ibid.). - To draw this conclusion and to pay heed to its normative significance would be a modicum against the very common mistake of taking what is in fact a parochial western value of individual decisional autonomy for an allegedly universally cherished moral value.

A more adequate view of autonomy would attend to the communal component inherent in, but hidden in standard analyses of, individual („personal") auton-omy. The communal component in individual autonomy can be grasped by con-sidering the fact that the ability to arrive at valid choices, a principal goal of auton-omy, implies being able to justify one's decisions in terms of reasons which not only oneself can accept but which one rightly (or, perhaps, wrongly) expects rele-vant others to find acceptable too. The decision of a client to undergo prophylactic breast amputation may be uncoerced and authentic and yet be based on reasons that would have very little justificatory strength with relevant others, i.e., other persons concerned by having some stake in the matter. Since most private choices in the workspace of genetic counseling have public consequences (e.g., reaffirm-ing and advancing the medico-legal arrangements and tendencies that have re-sulted in the client's ability to make these choices) it would be a flawed under-standing of „private choice" to think that for such choice there simply can be no relevant others beyond the agent her- or himself (cf. Mieth 1993). Genetic counsel-ing in matters of reproductive choice is in the majority of cases a relationship of the counselor with a couple, not a transaction with a single person. Moral theoriz-ing about genetic counseling is on the wrong track as long as it is perfused with an atomistic social ontology. It is in reaction to this shortcoming of the received view of autonomy (and its corollary in the theory of counseling, „non-directiveness") that some feminist bioethicists are beginning to amend a flawed concept of indi-vidual autonomy with what they term *autoceonomy* (Tong 1997, p.94), emphasiz-ing self ('auto') *and* community ('koinonia'). Autonomy in the new sense (more in line, I would argue, with the new genetics) incorporates a richer notion of know-ing-what-one-does, richer in the sense that agents have, or come to have with the aid of the counselor and other significant others, an understanding of the reasons for which they reject other options that are decisionally available to them, and a trust - achieved through reasoning, controversial if need be, with others - that their decisions have reasons behind them that are good enough to make the cho-sen way of action one with regard to which no significant other concerned by it in some relevant way would have grounds for legitimate objections.

The focus of moral concern, then, shifts away from respecting some person's or party's autonomy to procedures for *reconciling different autonomies* in morally responsible ways, that is in ways each party can accept without seriously compro-mising their moral point of view. Evidently, moral discourse as specified in dis-course ethics by reference to five parameters of a moral discourse (sec. III) pro-vides such a procedure. By 'different autonomies' I mean not only the (relative) decisional autonomies of different people. Rather, I mean the *ability* (of one or more interacting parties) *to intentionally frame and advance through practical de-liberation an agenda of one's own, within the scope of what one takes oneself, and is taken by others, to be responsible for.*

Naturally, a determinate agenda of one's own will vary with the various responsibilities that one bears in the matter at hand and with the priority ordering that one imposes on these responsibilities. Counselors, for instance, bear (amongst other things) a responsibility regarding the counseling process in virtue of their representing a community of co-professionals. At the same time, they bear certain responsibilities regarding the counseling process by virtue of their citizen status (e.g., a responsibility to obey the relevant laws); and some might bear certain other responsibilities in virtue of, say, their religious affiliation (e.g., a responsibility not to consider abortion, though legal, a morally right choice). Clients, for their part, have to fill out autonomously various responsibilities which they bear in virtue of, e.g., having been constituted (by their legal system) as legal subjects of reproductive freedom; as (would-be) parents; as partners (marital or otherwise); as members of some particular welfare system, etc. (cf. Kettner 1998a).

The conceptual division that has also played a considerable role in this conference, between *consumers* of genetic services („consumer values") and *providers* („provider values") in no way exhausts the space of different autonomies that need to be reconciled in the micro-context of good genetic counseling and in the macro-context of the profession's political representation within health related public policy making (Kettner & Schäfer 1998d). Extant differences in autonomies (within one person/party or between persons/parties) become *morally* relevant differences when their respective construals of pertinent facts, values and norms can be seen to make a difference in their interpretation of at least one of the five parameters of a moral discourse. The key *topoi* within the received view of autonomy - competence and informedness, non-coercion, and authenticity - map specifically into parameters one („reasonable articulation of need-claims"), two („bracketing of power differentials"), and three („non-strategic transparency") respectively. However, the framework of discourse ethics makes it clear that what matters is not solely the autonomy of the client but the autonomy of the counselor too, and furthermore, via parameter five („comprehensive inclusion"), the autonomy of everyone with a stake in a specific type of decision that comes to be made. In addition, parameter four („fusion of moral horizons") helps to draw a line between toleration of views that are morally deviant from what oneself believes to be morally right or wrong in the matter, and sheer indifference to such views under the guise of „value neutrality".

Should we stop counting as morally relevant the difference between someone's intention to counsel directively and someone else's intention to do so nondirectively?[11] Not at all. But we should rethink our answers to the question, What is it about directiveness and non-directiveness that allegedly makes the for-

---

[11] I refer to „intentions" here rather than to actions since to what extent it is possible to be true to those intentions in practice is a different (though related) question. Fine-grained empirical research on this question is now coming forward (cf. Hartog 1996).

mer morally right and the latter morally wrong in the context of genetic counseling? The promising answers, I suggest, are to be found in the following direction:

Both the counselor and the client(s) bring agendas of their own into the counseling process. For a number of reasons which have mostly to do with professionalism and the institutional arrangements in which genetic counseling processes normally take place[12] the counselor tends to be systematically advantaged in terms of (various sorts of) power inherent in such processes. *Nondirectiveness* (in the new reading that I suggest we should give this notion) is a moral responsibility that is incurred by counselors by virtue of their quality as professionals: a professional moral responsibility. The content of this professional moral responsibility is *to see to it that one does not use one's power advantage in the interaction with the client in order to advance one's own agenda on cost of the client's agenda.* A counselor's directiveness, in turn, is just that: using one's systematic power advantage in the interaction with the client to the effect of advancing one's own agenda on cost of the client's own agenda. When helping someone to make personally valid decisions is the morally right goal, directiveness in the explained sense is morally wrong if consequently the client's agenda will turn out to be less well represented in the client's practical deliberation and final decision than would otherwise be the case, thus alienating the client from his or her stance and decision. So it is essentially the „on cost of the client's agenda" bit in the formulation that I have suggested which introduces into the very definition of directiveness a normative element of prima facie moral wrongness, given the goal that counseling ought to enhance, or at least ought not to diminish, a client's autonomy.

Conclusion:
Human Rights as a Frame for Moral Discourse on the New Genetics

I conclude with a remark on the shape of applied ethics with regard to complex processes of factual, evaluative and normative change, as witnessed by the impact of the new genetics on already entrenched social practices relating to genetics.

Applied ethics does not inject normativity into a field of practices that is encountered as a normative void. There are no normative voids in human forms of social life. It would be more adequate to conceive of applied ethics as an attempt to intervene with certain moral resources of reflective normative governance into the multifarious normative governance processes, non-moral and moral, which go on all the time in any targeted field of practices. Applied ethics has to tap moral resources that are already *in* the practices it seeks to morally ameliorate. It operates from within, internal to the practices, otherwise it will invariable appear in the

---

[12] For an empirically rich comparison of genetic services within Europe, see Harris, R., et al. (1997).

roles of an invader, colonizer, or despot, roles which under normal circumstances virtually guarantee that not much good will come of such „moralizing" efforts.

It seems clear, to the participants of this conference at least, that the globalizing nature of the new genetic knowledge, technologies, and corresponding social practices („genetic services") calls for *global* regulatory efforts. From the point of view of an already established profession, such as human genetics as a medical specialty, it is desirable that such global regulatory efforts be as much self-regulatory as possible rather than externally imposed (e.g., by state law). This point is forcefully made by Dorothy C. Wertz and John C. Fletcher.[13] For professional self-regulation to reproduce the trust that society invests in some group of experts when a space for collective autonomy is accorded to this group by the rest of society, thus recognizing the group as a profession, the emerging profession's activities and goals must not be beset with too much serious moral perplexity. Otherwise, the profession will fall into disrepute and become subject to rigid external review and regulation; it will lose social credit and eventually perhaps even its recognition as a profession. Today, the new genetics appears on the mass-mediated scene of public discussion as deeply ambivalent, an object of utopian desires and of deep suspicions as well (Niccols 1998; Silver 1997). This situation renders applied ethics a necessity relative to the objective interest of the profession in maintaining its professional autonomy uncurtailed. The new genetics calls for an applied ethics that is suited to its moral problems and matches the scope of the perplexity which they arouse world-wide. But where should applied ethics seek its unifying frame?

One is tempted to say that answers will vary drastically depending on the kind of issue that is at stake in particular endeavors of applied ethics. There will be many different unifying frames across many different realms of problems and many different subspecialties of applied ethics. Why should the ethics of public accounting, for instance, share its unifying frame with, say, the ethics of animal liberation, or with medical ethics? Its grain of truth notwithstanding, this „balcanizing" impulse in applied ethics had better be resisted. However, its grain of truth is that it looks not very promising to expect all branches of applied ethics to be narrowly governed by a single small set of morally normative contents, be it in the format of moral values, moral rules, or moral principles. Maybe a helpful analogy here is to compare different branches of applied ethics with different states of law, each of which has its own particular constitution and characteristic tradition of law, none of which could be reduced to one super-constitution though each has assimilated, and embodies, the normative contents of human rights law, above all the General Declaration of Human Rights and the variously normative textures that are its offspring.

---

[13] See their contribution to this volume.

To pursue this analogy slightly further, I want to suggest that we think of applied ethics as a social reform movement that is in the process of globalizing itself into a world-wide network, mostly from „below" (think of the diffusion of Kennedy Institute Principlism in many bioethics initiatives and educative ethics committees in a great many corners of the world) but also from above (think of the UN bioethics committee). For understanding the unifying organizational paradigm of applied ethics, then, one would have to look, not to parliaments, state legislatures, engineering corporations, or academic scientific disciplines, but to non-governmental organizations of civic minded people, mostly professionals, decentralized but associated in the common pursuit of issues and interests that are perceived to have a salient connection with the public interest.

Considering the global outbreak of the applied ethics movement, and taking into account its resemblance to civil-society-oriented nongovernmental organizations, I suggest that the proper unifying frame for applied ethics, at least in cases where the targeted practices are themselves of a global nature, should be the normative texture of human rights and related textures of globalized normative import. Yet human rights are seriously undertheorized in the theory of the practice of applied ethics. Centering applied ethics on a moral theory of human rights would have a number of advantages over centering applied ethics on less entrenched, and less globally realized normative textures (such as different national professional codes of ethics, or different sets or hierarchies of „bioethical principles"). I am not certainly not advising to forget bioethical principles, or grand moral theories (such as Kantianism, Utilitarianism, Contractualism), and to substitute human rights for them.[14] Rather, what I am suggesting is that we look for a better theoretical integration of parochial principles, grand moral theories, and globalized morally relevant normative textures like those of human rights. This will considerably strengthen the practical purport that is at the heart of applied ethics. And it will allay (though not remove) skeptical criticism as to the colonizing nature of applied ethics, specifically bioethics, in non-western sociocultural contexts.

Unleashing the transformative power of applied ethics on the new genetics will effect marked changes in a motley set of normative textures, such as WHO guide-lines, national guide-lines, national bio-law, professional codes of ethics, curricula of professional training, „good clinical practice" standards, not to forget health care competence in the population fostered by schools, institutions of public learning, and mass media programming. Again, in order for such diverse modifications to cohere and work in the same direction rather than disconcert and defract, no other morally relevant steering framework appears to be better suited than the framework of human rights. Since this framework is already recognized

---

[14] I have argued elsewhere (Kettner 1992) that applied ethics, contrary to what some „casuists", „situation ethicists", and communitarian „virtue ethicists" think, cannot separate itself from normative frameworks of basic moral theory, notwithstanding the abstractness and aloofness of such frameworks.

globally so that it is safe to speak of an already at least weakly globalized „human rights culture", this framework serves as an interface for applied ethics and positive law in different moral and legal cultures world-wide.

From a discourse-theoretical perspective on human rights, three articles of the Universal Declaration of Human Rights of 1948 bear a special structural significance.[15] Article 28 is the performative enactment of the generic form of human rights. This article postulates a world order in which it would be possible for everyone to live up to every more specific human right. Article 1, with its appeal to egalitarian human dignity, alludes to a foundation of all would-be human rights, a foundation that should be available across different cultures with perhaps drastically different moral horizons and legal cultures. Article 19, freedom of information, articulates a presupposition that is required whenever determinate contents are loaded into the generic form of human rights under the normative expectation that this particular specification of contents proceeds in ways that everyone concerned could accept.

Moreover, article 19 might serve to situate the social movement of applied ethics vis-à-vis extant realities of already globalized mass-mediated forms of communication: Achieving a presence, showing face in national as well as transnational fora of public debate is for applied ethics, specifically for bioethics, not a lesser affair, not a peripheral agenda, not an expendable goal; rather, it is the major site of its globalization from below. In order to achieve such a presence, the ethics of the new genetics will have to nest itself critically within public spheres of mass mediated communication. Establishing bioethics committees at national and international levels within administrative flows of power and information just is not sufficient for this task. Arguably, it is not even necessary.[16]

References

Apel KO (1987) The problem of philosophical foundations in light of a transcendental pragmatics of language. In: Baynes K & Bohman J & McCarthy T (eds) After philosophy. End or transformation? MIT Press, Cambridge, pp 250-290

Apel KO (1980) Towards a transformation of philosophy. Routledge, London

Apel KO (1989) Diskurs und Verantwortung. Suhrkamp, Frankfurt

Bartels D & LeRoy BS & McCarthy P & Caplan AL (1997) Nondirectiveness in genetic counseling: A survey of practitioners. Am J of Medical Genetics 72:172-179

---

[15] Articles 2 (nondiscrimination!), 12 (protection of an individual's privacy!), 16 (freedom in marriage, protection of the family!), and 27 (progress in science should benefit everyone!) speak more to the *content* of many morally perplexing issues that are characteristically associated with the new genetics. (For a standard overview of such issues, see already Wertz & Fletcher 1989, p.458-461; for more recent listings, see what has become standard fare under the title of *Ethical, Legal, and Social Issues* (ELSI).

[16] For a normative thematization of the mass media, see Kettner (1997).

Beauchamp TL & Childress J (1989) Principles of biomedical ethics. University Press, Oxford, 3rd edition

Benjamin M (1990) Splitting the Difference. Compromise and integrity in ethics and politics. University Press, Kansas

Bosk CL (1993) The workplace ideology of genetic counselors. In: Bartels DM & LeRoy BS & Caplan AL (eds) Prescribing our future. Ethical challenges in genetic counseling. Aldine de Ruyter, N.Y., pp 25-38

Caplan AL (1993) Neutrality is not morality: The ethics of genetic counseling. In: Bartels DM & LeRoy BS & Caplan AL (eds) Prescribing our future. Ethical challenges in genetic counseling. Aldine de Gruyter, N.Y., pp 149-167

Charlesworth M (1993) Bioethics in a liberal society. University Press, Cambridge

Cheuvront B & Sorenson JR & Callahan NP & Stearns SC & DeVellis BM (1998) Psychosocial and educational outcomes associated with home- and clinic-based pretest education and cystic fibrosis carrier testing among a population of at-risk relatives. Am J of Medical Genetics 75:461-468

Clouser KD (1996) Concerning the inadequacies of principlism. In: Gert et al, op cit, pp 57-76

Comings D (1980) Prenatal diagnosis and the „new genetics". Am J Hum Genet 32:453-454

Dewey J (1939) Theory of valuation. University Press, Chicago

Es R (1993) Negotiating ethics. Eburon, Delft (NL)

Gert B et al (eds) (1996) Morality and the new genetics. A guide for students and health care providers. Jones and Bartlett, Boston

Gert B (1988) Morality: A new justification of the moral rules. Oxford University Press, N.Y.

Gert B (1998) Morality. Its nature and justification. University Press, Oxford

Gert B et al (1997) Bioethics: A return to fundamentals. Oxford University Press, N.Y.

Habermas J (1993) Justification and application: Remarks on discourse ethics. MIT Press, Cambridge

Harris R et al (1997) Genetic services in Europe. A comparative study of 31 countries by the Concerted Action on Genetic Services in Europe. European Journal of Human Genetics 5(2)

Hartog J (1996) Das genetische Beratungsgespräch. Institutionalisierte Kommunikation zwischen Experten und Nicht-Experten. Narr, Tübingen

Jennings B (1999) Liberales Autonomieverständnis und bürgerliche Interdependenz. Politische Kontexte angewandter Ethik. In: Kettner M (ed) Angewandte Ethik als Politikum. Suhrkamp, Frankfurt, in press

Kettner M & Schäfer D (1997a) Moral concern over cryopreserved human embryos: too much or too little? Human Reproduction 12(1):10-11

Kettner M & Schäfer D (1998d) Identifying moral perplexity in reproduction medicine. A discourse ethics rationale. Human Reproduction and Genetic Ethics 4(1):8-17

Kettner M (1992) Drei Dilemmata angewandter Ethik. In: Apel KO & Kettner M (eds) Zur Anwendung der Diskursethik in Politik, Recht und Wissenschaft. Suhrkamp, Frankfurt, pp 9-28

Kettner M (1997) Mass media, ethics and democracy. A normative Bermuda-triangle? In: Koller P & Puhl K (eds) Current issues in political philosophy: justice in society and world order (pp 278-292). Hölder-Pichler-Tempsky, Wien

Kettner M (1998a) Zum Kontext von Beratung als Zwang. In: M. Kettner (ed) Beratung als Zwang. Schwangerschaftsabbruch, genetische Aufklärung und die Grenzen kommunikativer Vernunft. Campus, Frankfurt, pp 9-46

Kettner M (1998b) Reasons in a world of practices. A reconstruction of Frederick L. Will's theory of normative governance. In: Westphal K (ed) Pragmatism, reason, and norms. University of Illinois Press, Chicago, pp 255-296

Kettner M (1998c) Neue Perspektiven der Diskursethik. In: Grunwald A & Gethmann C.F. (eds) Ethik technischen Handelns. Praktische Relevanz und Legitimation. Springer, Heidelberg, in print

Kettner M (1999) „Communicative rationality", good reasons, and argumentation. In: Nida-Rümelin J. et al (eds) Rationality, realism, revision. Proceedings of the 3rd International

Conference of the German Society for Analytic Philosophy (GAP), Munich 1997. Springer, Berlin, in print

Kim J (1994) Concepts of supervenience. In: Kim J: Supervenience and mind. Selected philosophical essays. University Press, Cambridge, pp 53-78

Lippman-Hand A & Fraser C (1979) Genetic counseling - the postcounseling period: 1. Parents' perception of uncertainty. Am J of Medical Genetics 4:51-71

Michie S & Bron F & Bobrow M & Marteau T (1997) Nondirectiveness in genetic counseling: An empirical study. Am. J. Hum. Genet., 60, 40-47

Mieth, D (1993): The problem of „justified interests" in genome analysis. A socioethical approach. In: Haker H & Hearn R & Steigleder K (eds) Ethics of human genome analysis. Attempto, Tübingen, pp 272 -289

Niccols A (1998) director: GATTACA. (Movie Film)

Outka G & Reeders JP (1993) (eds) Prospects for a common morality. University Press, Princeton

Rehg W (1994) Insight and solidarity. The discourse ethics of Jürgen Habermas. University of California Press, Berkeley

Scholz C & Kroner W (1989) Amniocentesis, risks, and 'mhm's. In: Sikkens EH et al (eds) Psychosocial aspects of genetic counseling. Proceedings of the First European Meeting on Psychosocial Aspects of Genetic Counseling. 9-11 November 1988. Globe, Groningen, pp 34-51

Shiloh S & Sagi M (1989) Effect of framing on the perception of genetic recurrence risks. Am J of Medical Genetics 33:130-135

Silver LM (1997) Remaking eden. Cloning and beyond in a brave new world. Avon Books, N.Y.

Singer GS (1996) Clarifying the duties and goals of genetic counselors: Implications for nondirectiveness. In: Gert et al, op cit, pp 125-156

Taylor PW (1961) Normative discourse. Greenwood Press, Westport (CT)

Walters L (1993) Ethical obligations of genetic counselors. In: Bartels DM & LeRoy BS & Caplan AL (eds) Prescribing our future. Ethical challenges in genetic counseling. Aldine de Gruyter, N.Y., pp 131-148

Weatherall DJ (1985) The new genetics and clincial practice. University Press, Oxford

Wertz DC & Fletcher JC (1990) Ethics and human genetics. A cross-cultural perspective. Springer-Verlag, Berlin

Appendix

Guidelines on Ethical Issues in Medical Genetics and the Provision of Genetic Services

Dorothy C. Wertz

In July, 1995, the Hereditary Diseases Programme of the World Health Organization published a 91-page document entitled „Guidelines on Ethical Issues in Medical Genetics and the Provision of Genetic Services." This document, available free of charge from Dr. Victor Boulyjenkov of WHO, CH-1211 Geneva, 27, Switzerland, was the result of a February, 1995 consultation among the following persons.

Dorothy C. Wertz, PhD
Social Science, Ethics, and Law
The Shriver Center Waltham, MA, U.S.A.

John C. Fletcher, PhD
Center for Biomedical Ethics
University of Virginia
Charlottesville, VA, U.S.A.

Kåre Berg, PhD, MD
Institute of Medical Genetics
University of Oslo, Norway

in cooperation with

Victor Boulyjenkov, PhD, MD, Director
Hereditary Diseases Programme
Division of Noncommunicable Diseases
World Health Organization
CH-1211 Geneva 27, Switzerland

Dr. Wertz has a background in social science and social ethics, and has been doing surveys on ethics and genetics since 1984. Dr. Fletcher's background is in bioethics. Professor Berg is a practicing medical geneticist with a special research interest in heart disease, for which he counsels individuals and families at risk.

WHO undertook the writing of the document because genetics poses some unique ethical problems not found in other areas of medicine. These appear in Table 1.

Most genetics professionals work in developed nations. Table 2 shows the approximate members of genetics professionals in developed and developing nations, according to a survey by Wertz and Fletcher in 1993-96, that included all nations with ten or more practicing geneticists. Numbers of genetics professionals in the United States (including masters-level genetic counselors) have grown substantially since the survey.

The eight basic ethical problems in medical genetics appear in Table 3. The WHO Guidelines address all of these issues, especially the voluntary nature of genetics services, the need for nondirectiveness in genetic counseling, full disclosure to persons counseled, freedom of choice about reproductive options, and privacy from institutional third parties. The Guidelines developed in part as a foundation for a future international code of ethics in medical genetics.

Major arguments in favor of a code of ethics appear in Table 4. At this time in history, when public education about genetics and its uses in medicine and society is so woefully inadequate, even in developed nations, it is important that genetics services be provided in the most ethical climate possible, in order to protect peo-

Table 1. How genetics differs from other areas of medicine

Genetics

1. Provides information about other blood relatives.

2. Provides information predicting the future of persons who are now healthy, including children.

3. Provides unexpected nonmedical information, such as nonpaternity.

4. Has a history of „nondirectiveness" in counseling.

Table 2. Worldwide distribution of geneticists

	Geneticists	Population (approx.)
Developed nations	3291	733.928.000
Developing nations	989	3.574.133.000
Eastern European nations	590	409.042.000

Table 3. Eight ethical problems in medical genetics

1. Access to/demand for services
2. Abortion choices
3. Confidentiality problems
4. Protection of privacy from 3rd parties
   – insurers
   – employers
   – government agencies
5. Disclosure dilemmas
6. Indication for prenatal diagnosis
   – to benefit others (e.g. prenatal paternity testing; tissue-typing a fetus
     as a potential organ donor)
   – sex selection
7. Voluntary or mandatory screening
8. Non-directive counseling

Table 4. Arguments for a code of ethics

- Protect present and future patients
- Reduce the public's fear of genetics
- Prevent restrictive laws
- Prevent lawsuits
- Transmit moral experience to next generation
- Influence public policy
- Promote international cooperation
- Make genetics accountable to society
- Improve image of the profession
- Improve moral climate for research

ple and to reduce public fears about genetics. In the absence of a professional code, there will be a tendency of governments to ban particular services or areas of research that could be useful in alleviating suffering or developing future treatments. Once passed, laws are often difficult to overturn. An ethical code would help to prevent passage of restrictive laws. It could also help to prevent lawsuits, by preventing professionals from the kinds of practices that could give rise to lawsuits, such as withholding information from patients or breaching confidentiality.

A code of ethics is one of the hallmarks of a profession, which designates a new specialty as a unique profession rather than a branch of some already existing profession. A code of ethics transmits the moral experience of one generation of professionals to the next generation of professionals. This is particularly important in genetics, where half of today's professionals have entered the field within the last 12 years.

Table 5. Arguments against a code of ethics

- Ethics is to subjective
- Reproductive decisions are too personal
- Technology changes too fast
- Increase in laws and lawsuits
- Exclusion of minority ethical views
- Code might not be used
- Restriction of beneficial research

Genetics professionals would be in a better position to influence public policy if they had a code of ethics. A code could also promote international cooperation, by assuring professionals that their colleagues in other countries adhered to similar standards, for example, informed consent.

A code could make genetics more accountable to society and improve the moral climate for research, testing, and treatment. Although a code would improve the moral image of the profession, this should not be its major purpose.

The arguments against a code of ethics appear in Table 5. The first two arguments are that ethics is too subjective and is a matter of case-by-case decisions by individual professionals and persons; therefore no code is possible. The authors of the WHO guidelines agree that in medicine many decisions are made partly on the basis of individual circumstances and personal beliefs, but these decisions are made within a larger framework of principles and relationships such as would be embodied in a code.

The argument that „technology changes too fast" for a code of ethics to keep pace with it is false. New technologies arise, but the basic ethical problems listed in Table 3 remain the same. Basic underlying principles include social justice (access to genetic services), personal autonomy (rights to know, rights to decide, rights to confidentiality), avoidance of harm, protecting the integrity of the family, and increasing the Common Good (beneficence). New techniques such as multiplex tests, cloning, and germline therapy will not change the basics of the code, though international bodies such as WHO may wish to add to it from time to time.

For reasons described above, a code will not increase the number of laws and lawsuits. Exclusion of those whose views are in the minority is a more complex problem. Ideally, a code would be agreed upon only after hearing as many views as possible, including those of patients and parents, and would be broad enough to not to oppress anyone.

Probably the best argument against a code is that it will not be used. This is a real possibility in nations with a strong belief in the autonomy of medical practitioners and a tradition of economic free enterprise in medicine, such as the United States. In many nations, however, especially in developing nations, discussion of ethical issues in genetics services is just beginning, and public health officials are seeking the guidance that a code would provide.

A code could restrict potentially beneficial research if it tried to ban some avenues of research outright. The authors of the WHO Guidelines have tried to leave the door open to future research, while arguing that all research with humans must be preceded by adequate animal studies proving safety and efficacy, and must be approved by the appropriate ethical boards within a country.

A list of existing international codes of ethics appears in Table 6. None of these deals with genetics. The July, 1997 UNESCO „Declaration on The Human Genome" speaks of „the dignity of the human genome" but provides no guidelines whatsoever for the practice of medical genetics. Therefore it cannot really be considered a code of ethics.

The contents of the WHO guidelines appear in Table 8. The guidelines stress the importance of individual decision-making after receiving full and unbiased information, the importance of receiving full information before all genetic tests, the absence of coercion by professionals or the state, the importance of preventing stigmatization and discrimination, and equal access to all legal genetics services, regardless of ability to pay. The Guidelines also state that use of prenatal diagnosis for sex selection is unacceptable. A summary appears in Table 9.

Table 6. International codes

1. World Medical Association,	
– Declaration of Geneva	1948
– Amended in Sydney	1968
2. Nuremberg Code	1949
3. World Medical Association,	
– Declaration of Helsinki	1964
– Revised in Tokyo	1975
4. International Council of Nurses	
– Code for Nurses	1973

Table 7.

North American organizations with code of ethics
• American Medical Association
• Canadian College of Medical Geneticists
• National Society of Genetic Counselors

Specialties with guidelines on specific problems
• American Academy of Pediatrics
• American College of Obstetricians and Gynecologists
• American College of Physicians

Table 8. Contents of 1995 guidelines

Goals of genetics
Priorities in health care systems
Basic principles
– Respect for persons
– Preserving family integrity
Genetic counseling
Informed consent
Rights to referral
Duty to recontact
Screening and testing
Disclosure and confidentiality of test results
Presymptomatic and susceptibility testing
Testing children and adolescents
Behavioral genetics and stigmatization
Adoption
Prenatal diagnosis: societal effects
Prenatal diagnosis: optimal services
Abortion after prenatal diagnosis
Preimplantation diagnosis
Keeping genetically impaired newborns alive
Protection of embryos from environmental harm
Research issues
DNA banking and privacy
Gene therapy

Table 9. Summary of guidelines

1. Equal access to available services.

2. Nondirective counseling.

3. Voluntary rather than mandatory services (except for newborn screening that benefits newborn).

4. Full disclosure of clinically relevant information.

5. Confidentiality, except when information could avert serious genetic harm to family.

6. Privacy from employers, insurers, etc.

7. Prenatal diagnosis only for health of fetus.

8. Availability of and respect for choices, including abortion choices.

9. Guidelines apply equally to adopted children.

10. Research follows protocols and informed consent. Includes preimplantation diagnosis.

11. National review for gene therapy protocols

Table 10. Appendices

International Bibliography on Ethics and Genetics, represents 21 countries
Informed Listing of Laws, Regulations, and Guidelines on Genetic Screening/Testing and Other
Aspects of Genetics

Table 11. Changes required by WHO

WHO Judicial Affairs Committee asked for:
1. Congruence with Cairo Conference statement that „abortion should not be promoted
   as a means of birth control".
2. Change „abortion" to „genetic abortion" so that WHO does not appear to support social
   abortions.

Suggested changes rejected by the authors
1. Omit „The woman should have the final power of decision".
2. Omit embryo research

Appendices are listed in Table 10. The International Bibliography was limited to 5 or 6 publications per country. The authors hope that the International Bibliography will be expanded in a future revision, and invite interested parties from all countries to send publications to Dr. Boulyjenkov at WHO.

The WHO Judicial Affairs Committee asked for the changes listed in Table 11. The authors agreed that abortion should not be „promoted as a means of birth control" and that planned unavailability of effective contraceptives such as the birth control pill (an unavailability that increases the number of abortions) was morally wrong. The term „genetic abortion," required by WHO, turned out to be confusing and will be omitted from any subsequent revision.

The authors retained the statement that „the woman should have final power of decision" in reproductive matters. However, as a result of requests by several Muslim countries in the Middle East at a subsequent 1997 meeting, the statement will be modified to „the woman should be an important decision-maker" in a future revision.

WHO requested omission of human embryo research because the authors of the Guidelines believed that such research would be necessary in order to develop effective gene therapy, and WHO was hesitant about supporting embryo research. The authors' original statement was retained in the document.

The process by which the document was developed appears in Table 12. CORN stands for „Council of Regional Networks of Genetics Services" (U.S.A.), and CCMG stands for „Canadian College of Medical Geneticists", the major Canadian

Table 12. Process

Summer 1993	Victor Boulyjenkov (WHO) asks Fletcher and Wertz for guidelines useful in developing nations.
January-March 1994	First draft
July-December 1994	Review by experts from 12 countries WHO Expert Advisory Panel on Human Genetics (9 persons) 7 Additional reviewers CORN Ethics Committee CCMG Ethics and Public Policy Committee
January 1995	Second draft
February 1995	3-day meeting in Geneva (Wertz, Fletcher, Berg, Boulyjenkov, WHO Secretariat)
March-June 1995	Review by WHO Secretariat (7 persons), WHO Office of Legal Counsel, WHO Office of Publications
August 1995	Publication of Document
Spring 1996	Summary article in WHO Bulletin
August 1996	Discussion at Permanent Committee of International Congress of Human Genetics in Rio de Janeiro.
March 1996	Discussed by panel at American College of Medical Genetics meetings San Antonio, Texas. Reviewed at this meeting by Dr. Maimon Cohen, past President of American Society of Human Genetics.
Sept 1996-Nov 1997	Reviews requested from additional experts. 31 replies from 24 countries, including Japan Society of Human Genetics.
October 1997	Representatives of International Federation of Human Genetics Societies (European Society of Human Genetics, American Society of Human Genetics, Australasian Society of Human Genetics) meet in Baltimore, MD, U.S.A.. As a result of reviewing the WHO Guidelines, they decide to write their own guidelines in a shorter version.
November 1997	WHO Guidelines discussed in Fukui, Japan at WHO Assisted Satellite Symposium on Medical Genetic Services and Bioethics, following UNESCO Asian Bioethics Conference. Japanese journalists express belief that the Japanese public is not ready for WHO Guidelines because of inadequate education about genetics and its ethical problems
December 14-15, 1997	15 WHO expert advisors, from all WHO regions, including developed and developing nations, meet in Geneva to revise guidelines. A 16-page document, representing a 100% consensus (not a majority vote) among all those present is developed and will be released after final revisions.

professional body. In both cases, the organization's Ethics Committee reviewed the document. It was also reviewed in 1994 by representatives of CIOMS (Council for International Organizations of Medical Sciences) and of the Association of Clinical Cytogeneticists of the United Kingdom.

Table 12 also presents the course of events after the November, 1997 Fukui meeting. The December 15-16 meeting in Geneva developed a 16-page document representing consensus of all those present. Its basic recommendations are similar to those in the previous document, but with more nuanced wording and greater attention to points of view not present in the earlier document, especially Muslim, Asian, and African views. The 16-page document will be preceded by a 3-page executive summary and submitted to the WHO Board of Directors in May for potential approval as an official WHO document. In order to differentiate it from the 1995 document, it will be entitled „*Proposed* Ethical Guidelines in Medical Genetics..."

The 91-page 1995 document will be revised for use as a „Background Document" for the Proposed Guidelines. The Background Document will present various points of view on the issues and the rationale for courses of action, but will not necessarily represent consensus.

The WHO expert advisors hope that the proposed guidelines will become the nucleus of an international code, similar to the Helsinki Declaration. They also expect future meetings with ongoing dialogue.